Ernest Ingersoll

Friends worth Knowing

Glimpses of American Natural History

Ernest Ingersoll

Friends worth Knowing

Glimpses of American Natural History

ISBN/EAN: 9783337026554

Printed in Europe, USA, Canada, Australia, Japan

Cover: Foto ©berggeist007 / pixelio.de

More available books at **www.hansebooks.com**

FRIENDS WORTH KNOWING

Glimpses of American Natural History

By ERNEST INGERSOLL

ILLUSTRATED

Nature is an admirable school-mistress.—Thoreau

NEW YORK
HARPER & BROTHERS, FRANKLIN SQUARE
1881

Entered according to Act of Congress, in the year 1880, by
HARPER & BROTHERS,
In the Office of the Librarian of Congress, at Washington.

All rights reserved.

CONTENTS.

The author gladly acknowledges his obligations to the publishers of the periodicals named below, for courteously permitting him to reprint these essays from their pages, as follows:

		PAGE
IN A SNAILERY	*Scribner's Monthly.*	9
FIRST-COMERS	{ *Appletons' Journal.* *Lippincott's Magazine.* }	36
WILD MICE	*St. Nicholas.*	57
AN ORNITHOLOGICAL LECTURE	*Scribner's Monthly.*	85
OUR WINTER BIRDS	*Appletons' Journal.*	106
THE BUFFALO AND HIS FATE	*Popular Science Monthly.*	140
THE SONG-SPARROW	*The Field* (London).	171
CIVILIZING INFLUENCES	*Sunday Afternoon.*	182
HOW ANIMALS GET HOME	*Scribner's Monthly.*	199
A MIDSUMMER PRINCE	*Harper's Magazine.*	222
BANK-SWALLOWS	*Popular Science Monthly.*	241

ILLUSTRATIONS.

	PAGE
THE MOUSE IN THE BIRD'S NEST	*Frontispiece*
BULIMUS, CYCLOSTOMA, AND OTHER TROPICAL SNAILS	11
ANATOMY OF THE COMMON WHITE-LIPPED HELIX	14
THE APPLE-SNAIL, AND ITS EGGS	16
THE COIL-SHELL (PLANORBIS) AND A LIMNEA	17
THE UNDER SIDE OF A WET CHIP	18
THE SNAILS OF THE TORRENTS	19
AN EDIBLE SNAIL	24
HELICES IN HUMBLE CIRCUMSTANCES	26
AN ALIEN IN THE CELLAR	28
HOUSE-WREN	40
THE HOUSE-MOUSE	58
THE JUMPING-MOUSE (JACULUS)	61
THE WHITE-FOOTED MOUSE	67
THE MOUSE AND THE OYSTER	70
LEAVING HOME	75
THE FIGHT WITH THE SNAKE	78
HAUNT OF THE HERON	86
THE KINGFISHER	87
SUMMER YELLOW-BIRDS	89
YELLOW-BREASTED CHATS	92
A JUNE MORNING	95
THE HUMMING-BIRD'S NEST	97
"ONLY A CAT-BIRD"	98

ILLUSTRATIONS.

	PAGE
Eagles	100
The Plover	101
The Wood-pewee	103
Turkey-buzzards	104
Snow-bunting	109
Brown Creeper	114
Cardinal-grossbeak	116
A Yellow-bird in Winter Dress	117
Crossbill	124
The Waxwing	130
Unwelcome!	132
A Shrike	139
A Mother and Child of the Plains	141
Travelling Herds	145
The Signal—Buffalo Herd in Sight	149
Indians Killing Buffaloes on the Upper Missouri	155
A Fight against Fate	159
The Bitter End	167
Fun for the Boys, but—	243
'A Narrow Escape	247

FRIENDS WORTH KNOWING.

I.

IN A SNAILERY.

Two-thirds of the persons to whom I show the little land and fresh-water mollusks in my snailery either start back with an "Oh! the horrid things!" which causes me some amusement, or else gaze straight out of the window, saying languidly, "How interesting!" which hurts my pride. I confess, therefore, that it is contrary to experience to attempt to interest general readers with an account of

"Ye little snails, with slippery tails,
Who noiselessly travel across my gravel."

Yet why not? Snails are of vast multitude and variety, ancient race, graceful form, dignified manners, industrious habits and gustatory excellence; *quod est demonstrandum.*

Snails differ from other gasteropodous mollusks chiefly

in that they are provided with lungs, and thereby are fitted to live in air instead of water. Hence all true snails are terrestrial. As the snail crawls upon a cabbage-leaf, all that you can see of the body is the square head bearing two long and two short horns, with the muscular base tapering behind. There is an oily skin, and on the back is borne a shell containing the rest of the body, twisted up in its spiral chamber. Extending along the whole under surface of the body is the tough corrugated disk upon which the animal creeps. This foot is the last part of the body to be withdrawn into the shell, and to its end, in a large division of pulmonate as well as marine mollusks, is attached a little horny valve which just fits the aperture of the shell and completely stops it up when the animal is within. This is called the operculum. The foot secretes a viscid fluid which greatly facilitates exertion by lubricating the path, and snails may often be traced to their hiding-places by a silvery trail of dried slime. So tenacious is this exudation that some species can hang in mid-air by spinning out a mucous thread; but, unlike the spider, have not the power to retrace their way by reeling in the gossamer cable. The slime also serves the naked species as a protection, birds and animals disliking the sticky, disgusting fluid; and it serves others as a weapon, seeming to benumb whatever

BULIMUS, CYCLOSTOMA, AND OTHER TROPICAL SNAILS.

small creature it touches. The oleacina, of Cuba, thus frequently is able to feed upon mollusks of twice its strength.

The snail possesses an elaborate anatomy for the performance of all the functions of digestion, respiration, circulation, and reproduction. A collar of nervous matter encircles the throat, whence two trunks carry nerves throughout the body, and filaments pass forward to the "horns," the longer and superior pair of which end in minute eyes and are called "eye-stalks," while the shorter pair are only tactile organs, and hence "feelers." These tentacles are as expressive as a mule's ears, giving an appearance of listless enjoyment when they hang down, and an immense alertness if they are rigid, as happens when the snail is on a march. The eyes are of little real use, being excelled for service by the senses of smell and taste, and it is doubtful whether the nerves generally are very sensitive, since a slug will be eaten without manifesting pain.

It is not surprising, perhaps, to find great tenacity of life in so lowly an animal; but Spallanzani, whose experiments with bats are celebrated, was the first to ascertain that not only parts of the head, but even the whole head might be reproduced, although not always. The shell is easily and frequently repaired, albeit hastily and not with the fine workmanship of the original.

14   FRIENDS WORTH KNOWING.

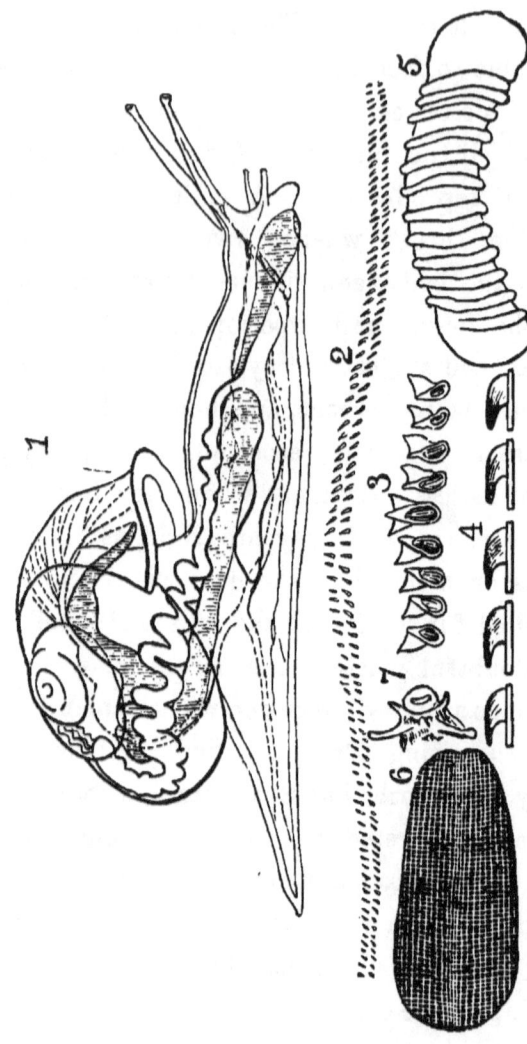

ANATOMY OF THE COMMON WHITE-LIPPED HELIX.

1. *a a*, eyes—*b b*, eye-stalks—*f*, foot—*m*, mouth; 2, a double row of teeth; 3, teeth highly magnified; 4, same—side-view; 5, jaw; 6, tongue showing the surface covered with rows of teeth; 7, mouth.

The pulmonates unite both sexes in one individual, but it requires the mutual union of two individuals to fertilize the eggs. The eggs are laid in May or June, when large numbers of snails gather in sunny places. When about to lay, the snail burrows into damp soil or decaying leaves, underneath a log, or in some other spot sheltered from the sun's rays, and there drops a cluster of thirty to fifty eggs looking like homeopathic pills. Three or four such deposits are made, and abandoned. This is the ordinary method of the genus helix, but some of the land and all the pond-snails present variations. The ova of slugs are attached by the ends in strings, like a rosary, and many deposits are made during the year. Bulimus and other South American genera isolate each egg, which in the case of some of the largest species is as big as a pigeon's. Vitrina and succinea glue them in masses upon stones and the stems of plants, while the tropical bulimi cement the leaves of trees together to form nests for their progeny. The pond-snails hang little globules of transparent gelatine containing a few eggs, or otherwise secure their fry to wet stones, floating chips, and the leaves of aquatic plants. In neritina, a brackish water inhabitant, the ova, immediately upon being laid, become attached to the surface of the parent's shell, and when the embryo hatches each egg splits about the middle, the upper part lifting off like a lid. Lastly, the eggs of the stout paludinæ of our

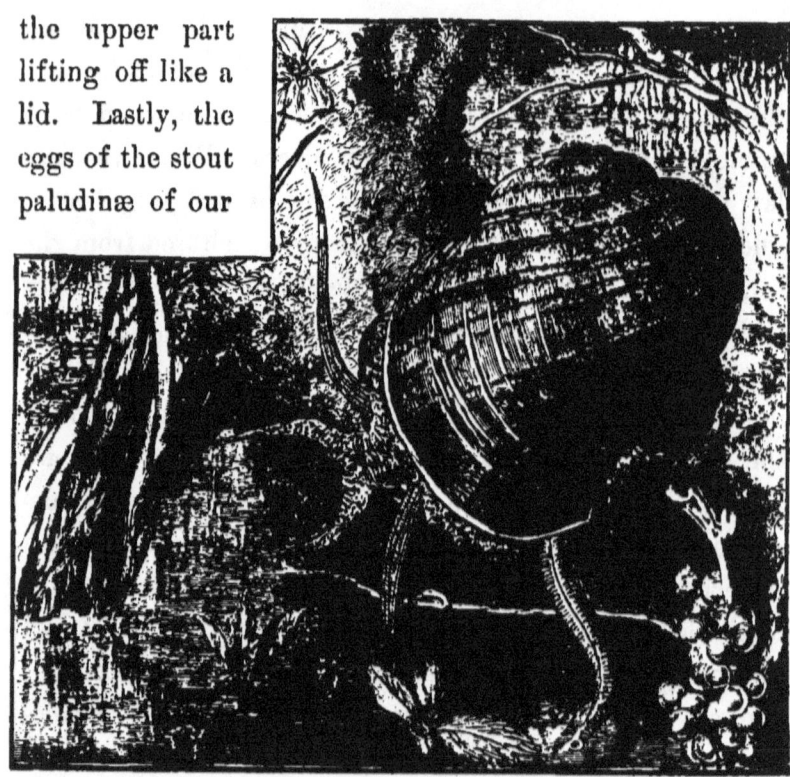

THE APPLE-SNAIL, AND ITS EGGS.

western lakes and rivers are not laid at all, but the embryos hatch out in the oviduct.

Under the microscope the translucent egg-envelopes present a beautiful appearance, being studded with glistening crystals of lime, so that the infant within seems to wear a gown embroidered with diamonds. Ordinarily the young snail gnaws his way out in about twenty or thirty days after the laying of the egg; but eggs laid in the autumn often remain unchanged until spring; and, indeed, may keep many years if they remain cool or dry. The vitality of snails' eggs almost passes belief. They have been so

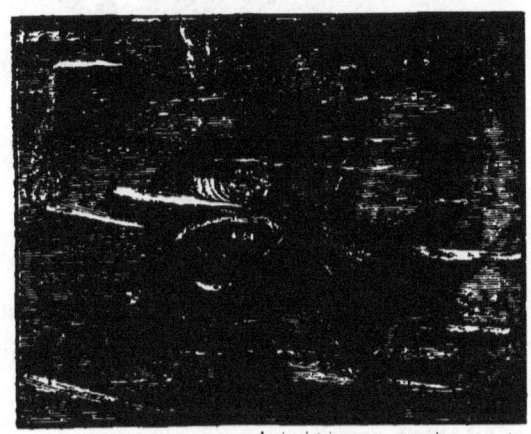

THE COIL-SHELL (PLANORBIS) AND A LIMNÆA.

completely dried as to be friable between the fingers, and desiccated in a furnace until reduced to almost invisible minuteness, yet always have regained their original bulk upon exposure to damp, and the young have been developed with the same success as from eggs not handled.

More or less wholly dependent on moisture, the young

snails at once seek out their habitual solitary retreats, and must be looked for under leaves, logs, and loose stones in

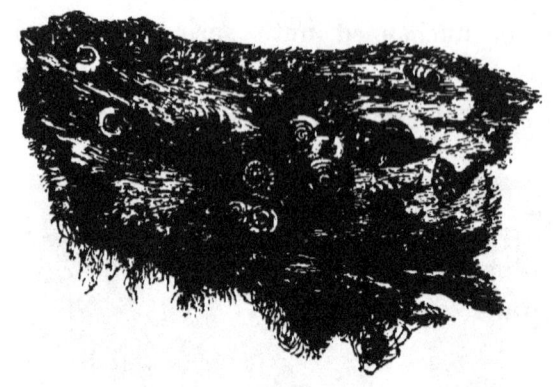

THE UNDER SIDE OF A WET CHIP.

the woods and pastures; at the roots of fern-tufts, lurking in the moss beside mountain brooklets; hiding in the crevices of rocky banks and old walls, crawling over the mud at the edge of swampy pools, creeping in and out of the crannies of bark on aged trees, or clinging to the under side of the leaves. Some forms are so minute that they would not hide the letter *o* in this print, yet you will soon come to perceive them amid the grains of mud adhering to the lower side of a soaked chip.

For fresh-water species, various resorts are to be searched. Go to the torrents with rocky bottoms for the paludinas and periwinkles (Melania); to quiet brooks for physas and coil-shells; for limneas to the reeking swamps and stagnant pools in the wet ooze. I know no better place in the world for pond-snails than the tule marshes of the Pacific slope, where hundreds of the great graceful *Limnea stagnalis* lie among the rotting vegetation, or float upside down at the surface of the still water. But some of the fresh-water mollusks remain most of the time at the bottom, coming to the surface only to breathe now and then; and to get their shells it is necessary to use a sieve-bottomed dipper, or

THE SNAILS OF THE TORRENTS.

some sort of dredge. When the water becomes low they bury themselves in the mud; it is therefore always profitable, late in the summer, to rake out the bottom of mud-holes where the water has entirely disappeared. Another plan is gently to pull up the water-weeds by the roots, and cleanse them in a basin of water. You will thus secure many very small species. Experience will quickly teach the collector where he may expect to find this and that kind, and that some caution and much sharpness of observation are necessary, since some species by their naturally dead tints, and others by a coating of mud, assimilate themselves so nearly to their surroundings as easily to be overlooked by man as well as other enemies.

The shell is increased rapidly for the first two or three years, and the delicate lines of increment, parallel with the outlines of the aperture, are readily visible on all the larger specimens. Various other signs indicate youth or adult age in the shell.

Mollusks prosper best, *cæteris paribus,* in a broken landscape, with plenty of lime in the soil. The reason, no doubt, why the West India Islands, the Cumberland Mountains, and similar regions are so peculiarly rich in shells of every sort, is that a ravine-cut surface and a wide area of limestone rocks characterize those districts; on the other hand, it is not surprising that I found nine-tenths of the Rocky Mountain species to be minute, since the geology is represented by sandstone and volcanic rocks.* Hot springs are very likely to be inhabited by mollusks, even when the temperature exceeds 100° Fahr., and the waters are very strongly impregnated with mineral salts.

Snails are mainly vegetarians, and all their mouth-parts and digestive organs are fitted for this diet. Just beneath the lower tentacles is the mouth, having on the upper lip a crescent-shaped jaw of horny texture, with a knife-like, or sometimes saw-like, cutting-edge. The lower lip has nothing of this kind, but in precisely the same attitude as our tongue is arranged a lingual membrane, long, narrow and cartilaginous, which may be brought up against the cutting-edge of the upper jaw. This "tongue" is studded with rows of infinitesimal silicious "teeth," 11,000 of which are possessed by our common white-lipped helix, although its ribbon is not a quarter of an inch long. All these sharp denticles point backward, so that the tongue acts not only as a rasp, but takes a firm hold upon the food. On holding the more transparent snails up to the light it is easy to see how they eat, and you can hear a nipping noise as the semicircular piece is bitten out of the leaf. Their voracity

---

* See Dr. Hayden's Report of the United States Geological Survey, 1874; and the Popular Science Monthly, July, 1875.

often causes immense devastation, particularly in England, where the great gray slugs will ruin a garden in one night, if the gardener is not daily on the watch. Our own strawberries sometimes suffer, but a border of sawdust, sand, or ashes around the bed is an adequate protection in dry weather. In trying to cross it, the marauders become so entangled in the particles adhering to their slimy bodies that they exhaust themselves in the attempt to get free. They also are very fond of fungi, including many poisonous kinds.

At the first hint of frost our snail feels the approach of a resistless lassitude, and, creeping under some mouldering log, or half-buried bowlder, it attaches itself, aperture upward, by exuding a little glue, and settles itself for a season of hibernating sleep. Withdrawing into the shell, the animal throws across the aperture a film of slimy mucus, which hardens as tight as a miniature drum-head. As the weather becomes colder, the creature draws itself a little farther in, and makes another "epiphragm," and so on until often five or six protect the animal sleeping snugly coiled in the deepest recesses of his domicile.

This state of torpidity is so profound that all the ordinary functions of the body cease—respiration being so entirely suspended that chemical tests are said to discover no change from its original purity in the air within the epiphragm. Thus the snail can pass without exhaustion the long cold months of the north, when it would be impossible for it to secure its customary food. This privilege of privacy reminds me of an old distich about another hibernater:

"The tortoise securely from danger does well,
When he tucks up his head and his tail in his shell."

The reviving sun of spring first interrupts this deep slumber, and the period of awakening is therefore delayed with the season, according to the varying natures of the different species. A few species, however, seem to hibernate very little, vitrina, for example, having been seen active in cold weather, and even crawling about in the snow; while the finest American specimens live high up on the Rocky

Mountains. At any time, nevertheless, an artificial raising of the temperature breaks the torpor, the warmth of the hand being enough to set the heart beating. Extreme drouth also will cause snails to seal their doors hermetically, without even hanging a card-basket outside. This is to shut off the evaporation of their bodily moisture, and happens in midsummer; hence it is termed æstivation. Cer-

AN EDIBLE SNAIL.

tain foreign slugs (Testacellidæ) which have no shells, are able to protect themselves under the same circumstances by a gelatinous appendage of the mantle, which, in case of sudden change of temperature, can be extended like an outer mantle, so to speak, from its place of storage, under the "buckler," and having wrapped themselves, they burrow into the soil. These carnivorous testacelles are the fiercest of all their race, and one might be excused for quoting:

> "But he lay like a warrior taking his rest
> With his martial cloak around him."

Snails are found in the most barren deserts and on the smallest islands all over the globe, reaching to near the line of perpetual snow on mountains, and restricted only by the arctic boundary of vegetation. There is a great difference between the snails of the tropics and those of high latitudes—size, number of species in a given district, and intensity of color decreasing as you go away from the equator; but this statement must be taken in a very general sense.* Different quarters of the globe are characterized by special groups of land mollusks as of other animals—thus, achatinella, with 300 species, is confined to the Sandwich Islands. But helix—the true snail—with its many subgenera and 2000 species, is absolutely cosmopolitan. The fresh-water forms, also, are spread everywhere, except in Australia, and flourish in cold countries, pupa having the hardihood to live

---

* Mr. A. R. Wallace's late work, "Tropical Nature," contained a long series of observations upon the colors of terrestrial mollusks among other animals. In two articles in "Science News," vol. i., pp. 52 and 84, Mr. Thomas Bland studied Wallace's principles in their application to American snails, and found that color is a matter of less account than it has hitherto been considered to be.

nearer the north pole than any other known shell. Yet it is a remarkable fact, that, however erratic and extensive may be the range of the genera to which they belong, the majority of the species of pulmonates of all sorts have an extremely limited habitat, in some cases comprising only a few square rods. A second noteworthy fact, obtaining in no other extensive group of animals, is, that many more species of land shells exist in the islands than on the continents of the world. Mr. A. R. Wallace accounts for this curious fact by explaining how certain influences make islands—particularly if long insulated

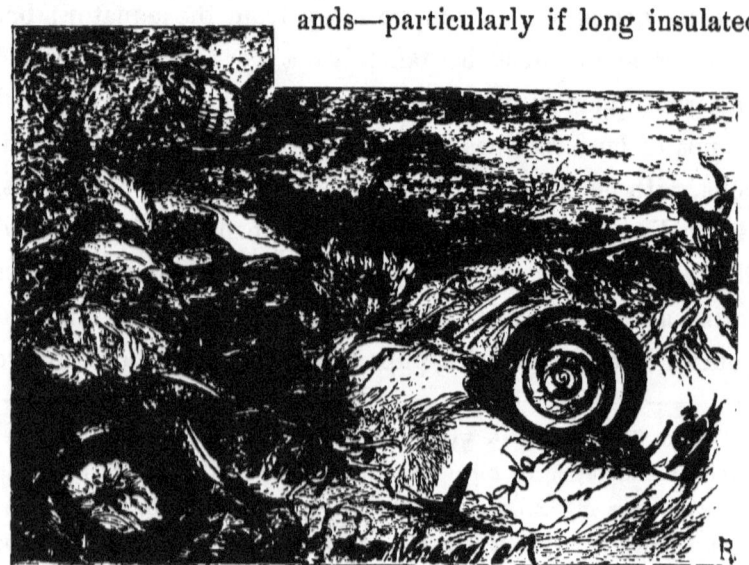

HELICES IN HUMBLE CIRCUMSTANCES.

—more productive than continents, and at the same time liable to be deficient in enemies to snails.

How has this curious distribution come to pass? How have seemingly impassable barriers been overcome, so that closely related forms are now found at the antipodes?

Snails are of domestic tastes. "The Heathen painted before the modest women's doors Venus sitting upon a snail, *quæ domi forta vocatur,* called a House bearer, to teach them to stay at home, and to carry their houses about with them." They are also slow of pace, as a list of poets are ready to stand up and testify; but they have had a long time in which to "get a good ready," first to start, and afterward to accomplish their travels, since their existence as a race goes back to when dark forests of ferns waved their heavy fronds over the inky palæozoic bogs. Distance disappears in the presence of such prodigious time. Lands like our western plains, now an arid waste impassable to mollusks, in by-gone ages were clothed with dense and limitless verdure, where every form of terrestrial life abounded. Between the present and even the laying down of those cretaceous sandstones that make the soil of our level plains, the Rocky Mountains have been elevated from an altitude at which any mollusk could probably have lived upon their summits, until now they may be a barrier to many species. Such changes may have happened anywhere, again and again, and thus the two halves of a community been divided. In succeeding centuries the members of the parted sections may have diverged in their de-

AN ALIEN IN THE CELLAR.

velopment, until on this side of a mountain range, or desert, or sea, we now find one set of species and on that side another set, which belong to the same genera, and may in some cases be proved, as well as surmised, to have had an identical origin.

But the main explanation of their dispersion is undoubtedly to be found in a land connection once existing between the different islands of present archipelagoes, and between these and the neighboring main-lands. It has been pretty satisfactorily demonstrated that during the glacial period the oceans must have been drained of water representing a universal depth of 1000 feet, in order to construct the enormously thick ice-caps which covered the polar hemispheres. This would expose a vast area of shallows, before and since deeply submerged, across which snails might easily migrate to other latitudes; when, at the end of the glacial period, the melted ice reclaimed the shallows, the snails would be left colonized upon the high points now widely separated by water.

More casual circumstances have always contributed to this world-wide distribution. Snails frequently conceal themselves in crevices of bark, or firmly attach themselves to branches and foliage, and thus might be drifted long distances, since they are able to resist starvation for an immense period, and protect themselves against injury from salt-water or excessive heat by means of opercula and epiphragms; violent storms might frequently transport living shells a considerable distance, aquatic birds carry them or their eggs from pond to pond attached to feet or plumage.

The astonishing vitality of the snails in every stage of existence favors the theory that they endure such accidental means of travel and thrive at the end of it. Professor Morse records that he has seen certain species frozen in solid blocks of ice, and afterward regain their activity; and enduring an equal extreme of heat, where the sun's rays crisped the leaves for weeks together, without any bad effect. They have been shut up for years in pill-boxes, glued for years (seven years in one case, Dr. Newcomb, of Cornell University, told me) to tablets in museums, and yet a trifle of moisture has been sufficient to resuscitate them. They survive so well being buried in the ballast of ships that at every seaport, almost, you may find species imported in that way, which came to life when the ballast was dumped at the time of unloading. That birds occasionally carry them about is well verified.

Such are some of the methods of dispersion. Yet students are obliged to confess that the causes of the present puzzling geographical distribution of land shells are so complex that we can hardly hope to determine them with much exactness.

As to the longevity of snails, little is known; but some individuals no doubt attain great age. Some species of cylindrella have a habit of deserting the point of the spire of their long, slender shells as they grow old. The abandoned portion speedily becomes dead, and cracks off upon the least injury. The sign of a perfect adult shell in these species, therefore, is that it is broken! Mr. Thomas Bland, the distinguished student of West Indian conchology, discovered this curious fact. After the cylindrella has thus voluntarily left the upper part of his shell, he builds a partition across behind him. Often other mollusks are driven to a similar expedient by accident or the decay of extreme age. This is called decortication, and is almost always to be seen in the beaks of the larger unios or fresh-water mussels of our inland rivers. The spiral shells most likely to be thus affected are those that live in swift running water, where the bottom is rocky—such as the members of the families viviparidæ and strepomatidæ. The latter are rarely seen otherwise than dreadfully broken.

Another curious thing is to be noticed in this connection: whole species sometimes suddenly die out. Not only a conchologist, but others, travelling through certain parts of our western territories, are struck by the prodigious quantities of dead white snail-shells scattered over the ground. These are the *Helix cooperi*, of which a few are still living in nooks and corners of the mountains. They are of all sizes, degrees of variation, and ages, and lie bleached in millions on the surface of the ground. Mr. E. L. Layard, in a recent number of the London *Field*, mentions a precisely similar case in Mozambique and another in Fiji. Why have these species thus suddenly become extinct? As far as we can see, there is no cause for their epidemic death.

Snails, being great eaters, meet their just reward in being eaten. The paludine forms are sought after by all sorts of water birds, particularly ducks and rails; while the thrushes and other birds crush the shells of the land snails and extract their juicy bodies. The woodland birds, however, will not eat the naked-bodied slugs: the slime sticks to their beaks and soils their feathers; but the ducks seem to have no such dainty prejudices. Some mammals, like the raccoons and wood-rats, also eat them; insects suck their juices, and the carnivorous slugs prey upon one another. Lastly, man, the greatest enemy of the brute creation, employs several species of snails for culinary purposes. By the Romans they were esteemed a great luxury, and portions of plantations were set apart for the cultivation of the large, edible *Helix pomatia*, where they were fattened by the thousand upon bran sodden in wine. From Italy this taste spread throughout the Old World, and colonies of this exotic species, survivors of classical "preserves," are yet found in Great Britain where the Roman encampments were. They are still regarded as a delicacy in Italy and France, the favorite method of preparation being to boil in milk, with plenteous seasoning. Frank Buckland says that several of the larger English species are excellent food for hungry people, and recommends them either boiled in milk, or, in winter, raw, after soaking for an hour in salt and water. Some of the French restaurants in London have them placed regularly upon their bills of fare. Thousands are collected annually and sent to London as food for cage-birds. Dr. Edward Gray stated, a few years ago, that immense quantities were shipped alive to the United States "as delicacies;" but I am inclined to consider this an exaggeration growing out of the fact that, among our fancy groceries "a few jars of pickled snails, imported from Italy," figure as a curiosity, rather than something needed, for the table. The same author records that the glassmen at Newcastle once a year have a snail feast, collecting the animals in the fields and hedges on the Sunday before.

Mr. W. G. Binney, for whom a sirup of snails was prescribed by two regular physicians in Paris in 1863, points out how old is the belief that land mollusks possess valuable medicinal qualities. In the Middle Ages the rudimentary shell of the slug acquired a high rank among the numerous bezoars and amulets which were supposed to protect the body from evil influences, and to impart health and activity. The accounts of these virtues, copied from one author to another, have perpetuated the early superstitions until it is difficult to overcome them by the light of the present day, when, even in England, snails are supposed to possess a useful power in cases of lung trouble. A full relation of all the absurdities which gained credence would form a curious and marvellous page in the history of credulity. They have, also, from very early times, been used in the preparation of cosmetics; and no longer than two or three centuries ago the water procured from them by distillation was much celebrated, and employed by ladies to impart whiteness and freshness to the complexion. Finally, I hear that there is celebrated in Rome, even now, a midsummer festival, upon which occasion all family feuds may be made up, or any differences between friends easily adjusted, since that is the spirit of the day; and a sign or token of this renewed friendship and good-will is the present of a snail from one party to the other, or an exchange of mollusks between them. The symbolism and virtue reside in the alleged amicable influence of the head and "horns"—why, perhaps comparative mythologists may be able to tell us.

In this country no such fanciful notions have ever gained credence. The snails are too habitually hidden to attract the attention of any but a few; and even when their existence is known, they are unfortunately regarded with such a disgust as would preclude any acceptance of them, either for food or medicine.

Yet why this disgust? Snails are of ancient race, vast variety, graceful shape, dignified bearing, industrious and peaceful habits, edible and curative properties; *quod erat demonstrandum.*

II.

FIRST-COMERS.

The lengthening of the days, as the year slowly advances, brings with it increased longing for still balmier weather to every one whose pleasure is not bound within the narrow limits of the opera and *soirée*. To the lover of long rambles in the woods and meadows, or of lazy boating along some placid stream, where the water-lilies bow to let him pass and buoyantly rise in his wake, shaking the drops from their shining fronds, every indication of approaching spring is eagerly scanned, and is hailed with delight. The slow decay of the ice in the ponds, the vivid green of the aquatic plants disclosed by its melting, the delicate herbage hiding under the sodden leaves, the gummy and bursting buds, all presage the charms of reviving nature. Then the sounds awake. The frogs bid each other good-morning after their long sleep; the lowing of calves and the bleating of lambs resound from the hill-sides; the tender warble of the bluebird, the cheery call of the robin, and the gurgle of swollen brooks, mingle in our ears as we pick our way along the muddy paths; until, some bright April morning, we discover that surly Winter is gone, and coy Spring is shyly waiting for us to bid her welcome.

In this company of the heralds of this admirable change of the seasons, none have a better part than the birds, whose wings bear beauty and song. Half a dozen of these messengers — the bluebird, the wren, the dove, and the blackbirds — are especially first-comers, and to them I ask attention. The song-sparrow also belongs here, by good right, but he enjoys an essay all to himself elsewhere.

Among the very earliest are the familiar bluebirds; indeed, they may occasionally be found all winter long in sunny fields. By All-fools-day they have become common, and are seeking their mates, which are soon found. Meanwhile, from every field, and about the yet desolate gardens, is heard the bluebird's cheery voice. It is a happy, contented warble, and, though no great credit belongs to the singer as a musician, his tender melody is among the most delightful of vernal sounds. There is an ubiquity or ventriloquistic peculiarity about this song—whether due to its quality or to the capricious breeze upon which it is usually borne, I do not know—which tends to make its source indefinite. You may hear the notes on a bright March morning, but cannot find their pretty author. He denies your eyes the welcome sight of him, until at last you give up the search only to discover him close behind you. This unintended ventriloquism may be in his favor, but his azure plumage is very conspicuous as he stands on a tall fence-rider with the woods for a background, or reconnoitres the entrance to an old woodpecker's hole in some white cottonwood, and many bluebirds are killed by the small hawks. Thoreau said that he carried the sky on his back, to which John Burroughs added, "and the earth on his breast." This describes him perfectly.

The bluebird is not ambitious in his flight, never emulating the lofty journeys of the pointed-winged birds, and is rarely seen sixty feet above the surface. He loiters about the outskirts of the woods, flitting from stump to stump; delights in a tract of newly-cleared land; and looks no farther when he discovers, not far from the farm-house, a group of charred and towering trunks—monuments of a long-passed fire in the forest. Next to that he loves an aged orchard. In both places the attraction is mainly the grubs, worms, and insects that infest dead and decaying woods, and upon which he feeds. To such a spot he leads his mate, easily to be distinguished by her duller plumage. Together they go house-hunting. It is not long, usually, before they are suited; for the woodpeckers have been there years before them, chiselling out many holes for themselves which are now left vacant; or the snapping off of some old limb has opened the way to a snug cavity in its hollow interior. Any kind of a cranny seems to serve in a pinch. I have known them to build in a broken tin water-spout under the eaves of a house for want of a better place; although, no doubt, the birds exercise a decided choice when they can. The tenement determined upon, the furnishing of it does not require much labor or contrivance. The birds bring enough of a peculiar kind of soft grass which turns reddish brown when it dries, sometimes mix with it a little hair, and thus thickly carpet the bottom of the cavity. That is all. The eggs are laid by the second week in April, and the young are hatched about ten days after. The eggs are five in number, and are light blue, without spots. Once, in Northern Ohio, I found a nestful of pearly-white eggs, and one other similar case has come to my knowledge. They were just as well worth sitting on, however, as five blue eggs would have been.

The bluebird is also a true bird of the garden, taking the place of England's robin-redbreast more nearly than any other bird in America. It is no trouble to have them twittering about the house the whole summer through. The

40   *FRIENDS WORTH KNOWING.*

negroes at the South always have an abundance of different birds about their cabins by simply hanging up empty

HOUSE-WREN.

gourds; and a cigar-box with a hole in it is all-sufficient. But you must not be disappointed if the house-wrens utterly dispossess the bluebirds of the houses you have put up, for thé wrens are regular buccaneers, with no more heart or conscience than a walnut; nevertheless, the bluebirds are far better fighters than one would suspect them to be, as the English sparrow has learned at the cost of many a sore spot.

This same house-wren is so well known that I need only allude to him; and any further description than to say that he is the wee brown bird, about as large as your thumb, which frequents the garden bird-boxes and the barn, is unnecessary. He comes early and stays late. He makes himself at home immediately, and is everywhere present, bustling about outhouses and barns, rapidly building his nest in the most insecure and unfrequented places, like the sleeve of an old coat left in the barn, or a lantern hung against the woodshed; and, if it is repeatedly pulled down, as often rebuilding it, literally "pitching into" other wrens, and bluebirds, and swallows, whom he considers trespassers on his right to the whole garden, and fighting so audaciously and persistently as nearly always to come off victor; squeaking in and out of every crevice, with his comical tail at half-cock; inquiring into every other living thing's business, yet not neglecting his own, this little bobbing bunch of brown excitement is the very spirit of impudence.

The wren does not confine himself altogether to the

garden, however. You may find him everywhere in the woods, and few species are equal to this in the number of individuals. An old stump that is too soft for the woodpeckers, or the hollow, broken limb of a tree that the winds have demolished, is his chosen home. Into a hole somewhere he stuffs a large quantity of twigs, some of them of astonishing size when we think how small a bird handles them. In the centre of this mass is a soft chamber, wherein six or seven brick-dust-colored eggs are hatched late in May. It is a nest which justifies his generic name, *Troglodytes*, and so fond of his queer den is he, and so restlessly active, that when his proper home is finished, he packs full of rubbish half the crevices in the vicinity, out of a sheer want of some better way to occupy his time and ease his energy.

There is one component of this nest which is also used by the vireos and gnatcatchers—namely, round pellets of a white cottony substance, the nature of which I was puzzled to determine. At last I caught the birds collecting it, and found it to be a minute fungus which covers dead twigs here and there with a living velvet of snowy white. It is elastic and somewhat viscous, and with gossamer serves an obvious purpose in such a nest as the vireo's; but why the wrens scatter it through their brush-pile is not so clear.

One of my pleasantest memories is of a sparkling April morning in 1874, at Scott's Landing, a little railway-junction on the Ohio River. It was bright and cold, and the wheezy steamboats passing up and down the river trailed from their tall and slender stacks great golden banners athwart the rising sun. The birds were up betimes. Crows from far and near were gathering to breakfast at the banks of the river, as is their custom at seasons of high water. The crow blackbirds—redundancy of title!—were moving in small flocks about some newly ploughed ground, smacking their horny lips at one another over some luscious, luckless grub; and their cousins, the military redwings, were in the highest glee. Cardinals are the natural bird-feature there; and their bold whistling resounded from every hill-side. Out of the orchard came the sharp squeak of a black-and-white creeper, the noisy chatter of chipping-sparrows, and the *dee-dee-dee* of the miniature Southern chickadees. One tree was the haunt of a single robin—*rara avis* in that locality —and he sang loud and long, not minding his loneliness. Bluebirds were not plenty, but a pair of them, and perhaps two families, inhabited an old cherry-tree so near to the railway-track that the tops of the passing cars pushed aside the boughs. I have noticed so many nests of birds built in close proximity to railways that I have thought the builders exercised a distinct preference for the situation, as making them safer from the attack of hawks.

Not an uncommon bird, hopping down between the rails to pick up the grain dropped from the freight-trains, was the turtle-dove, which was an old acquaintance of mine in the West, but which is rare in New England. They were very wary, uttered no note, and came with the silence of ghosts. If I only stirred when they were near, whir! away went my doves, straight and swift as an arrow, spreading their white-edged tails.

A portion of the following summer I spent on the Little Kanawha, and many a day was I entertained by the notes of the turtle-dove floating down from a hill-top as I threaded my way through the woods. Among the most common of birds in West Virginia, the people yet regarded it with affection, and made as great a disturbance if one was shot as they would at the shooting of a house-pigeon. They were jealous of the few purple martins they had in the same degree. Why it is called the turtle-dove I do not know. Probably because of its kinship with the turtle-dove of Europe; but this only puts the difficulty one step farther back. Its other name—mourning-dove—is more characteristic; for its song, if it may be called such, is a sobbing refrain, that, tolling from afar, recalls the echoing of a distant church-bell—

"Swinging slow with sullen roar."

The cry is frequently mistaken for that of some owl; but the dove does not sing at night, or some nervous people would grow wild. If it did, it would take character as a banshee, and become a bird of evil omen. On the contrary, its coming in early spring is now welcomed as one of the first signs of the sure advance of the season, and its plaintive note is only a minor-tone, mingling harmoniously with the livelier notes of other denizens of the woods.

The mourning-doves pair very early, and are as affectionate in their attachments as are most of the doves and pigeons, whose "billings and cooings" have become exaggerated into a proverb to express the first enthusiasm of young love. Their home is an indifferent affair, but perhaps its very scantiness may serve to benefit its owners by making it less conspicuous among the almost leafless branches, where it is likely to be placed early in the season. The nest is not by any means always in a tree, although a snug thorn-apple offers temptations that few doves can resist; but it may be put on the flat top of a stump, on the protruding end of a fence-rail, or the eggs may sometimes be laid on the ruins of a last year's nest, as in a case I once noticed where three dove's eggs were laid in an old cat-bird's nest, around the ruins of which the snow was yet unmelted. On the plains I have seen many times how these birds scratch a few grass-stalks together on the ground, for want of a better place. It is not to be wondered at that pigeons have been easily domesticated, when they accommodate themselves so readily to any exigency in rearing their young. However placed, this nest is a slight platform of twigs, just sufficient to hold the two or three eggs; or, if the top of a stump, or the ground, be chosen as the site, it is not uncommon to find simply a little rim, like a tinker's dam, built around the eggs, which themselves rest on the bare surface of the stump.

Another early and familiar visitor to the gardens is the chipping sparrow, or "chippy," its delicate voice coming to us from among the first blossoms of the lilac. It is also called the "hair-bird," because its nest is composed mainly of horse-hairs twined into a flat little basket of slender twigs and rootlets. But this is not a good name; the scientific designation, "social sparrow," fits the bird better, for it seeks to be social with man, and places its home where every boy and girl of the family may look in at the front door. The eggs are pea-green, scrawled, as though by a pen, with black lines and dots.

The food of the chippy during the spring and summer consists largely of small insects, and he searches carefully through the blossoming trees for the minute bugs that infest the leaves and flowers, occasionally nipping off the sweet and tender stamens of the apple and cherry blossoms, or taking wee bites out of the early currants. He flits quietly and busily all over the shrubbery, an image of a happy and contented little workman, *tra-la-la-ing* in a fine trilling voice, that would be shrill were it not so sweet, an aria from some bright bird-opera.

The chippy is so easily watched that I do not propose to tell all I have learned about it, and thus rob a reader of the pleasure of learning its beautiful ways for himself. You will not find it difficult to become acquainted with these pigmy sparrows after you have recognized their chestnut caps among your rose-bushes. You will see, also, that you may tame them and teach them to come to you for crumbs. They are almost the only birds that the insolent English sparrows will be friendly toward; and they are wonderfully devoted to their young: but I am forgetting that the reader was to find all this out for himself!

I have in mind the delta of a river whose shores are so

level that it is a constant struggle whether land or water shall prevail. The river finds its way to the broad harbor through a dozen or more channels, between which are low islands overgrown with great trees burdened and festooned with grape-vines and moss, and tangled with thickets and rank fern-brakes, or growths of wild rice and luxuriant water-weeds so dense and tall as to be impenetrable even to a canoe. Here blooms the magnificent lotus (*Nelumbium luteum*), with its corolla as large as your hat and its leaf half a boat-length broad—great banks of it, giving out a faint, sweet, soporific, almost intoxicating odor.

Curious sounds reach you as you thread the mazes of the swamp. The water boils up from the oozy bottom, and the bubbles break at the surface with a faint, lisping sound; the reeds softly rattle against one another like the rustle of heavy silks, and you can hear the lily-pads and deeply-anchored stems of the water-weeds rubbing against one another. More articulate noises strike your ear—the sharp-clucking lectures on propriety of the mud-hen to its young; the *brek-kek-kek, coaz-coaz* of the frog; the splash of a tumbling turtle; the rushing of a flock of startled ducks rising on swift wings; the sprightly, contagious laughter of those little elves, the marsh-wrens, teetering on the elastic leaves of the cat-tails.

Never absent from such a reedy picture are the blackbirds, especially the redwing (*Agelæus phœniceus*), whose favorite resort is where the rushes grow most densely, among which he places his nest. The little swales in the meadows, also, where tufts of rank grass flourish upon islands formed by the roots of many previous years' growth, and stunted alders and cranberry-bushes shade the black water, are nearly always sure to be the home of a few pairs, so that they become well known to everybody, whether inland or alongshore, as soon as the ice melts. Such extensive marshes as I have just described are, however, the great centres of blackbird population, where they breed, where they collect in great hordes of young and old as the end of the season approaches, and whence they repair to the neighboring fields of Indian-corn to tear open the husks and pick the succulent kernels. In September I have seen them literally in tens of thousands wheeling about the inundated wild-rice fields bounding the western end of Lake Erie, their black backs and gay red epaulets glistening in the sun "like an army with banners." The Canadian *voyageurs* call them "officer-birds," and the impression of an army before him is always strong upon the beholder as he gazes at these prodigious flocks in autumn. It is extremely interesting to watch the swift evolutions of their crowded ranks, and observe the regularity and concert of action which governs the movements of the splendidly uniformed birds.

The redwings are among the earliest of our vernal visitors, and south of the Ohio River and Washington may be found all through the winter. Their loud and rollicking spring note is one of the most invigorating sounds in nature, and most typical of the reviving year. *Conk-quirée! conk-quirée!* sings out the male, as though he knew a good story if only he had a mind to tell it; and then adds *chuck!* as though he thought it of no use to try to interest you in it, and that he had been indiscreet in betraying an enthusiasm beneath his dignity over a matter beyond your appreciation. His plain brown mate immediately says *chuck!* too, quite agreeing with her lord and master that it is not best to waste their confidence upon *you*.

The centre of all their interest is the compact, tight basket woven of wet grass-blades and split rush-leaves which is supported among the reeds or rests on a tussock of wire-grass surrounded by water. It is a model nest, and they understand so well the labor it cost that they are mightily jealous of harm coming to it. The eggs are five in number, of a faded blue tint, marbled, streaked and spotted with leather-color and black, in shape rather elongated and pointed. The fledglings are abroad about the 1st of June, when the parents proceed to the production of another brood.

These blackbirds have the bump of domesticity largely developed, and if their household is disturbed they make a terrible fuss, calling upon all nature to witness their sorrow and execrate the wretch that is violating their privacy.

During all the spring season, and particularly while the young are being provided for, the redwings subsist almost exclusively on worms, grubs, caterpillars, and a great variety of such sluggish insects, and their voracious larvæ, as do damage to the roots and early sprouts of whatever the farmer plants; nor do they abandon this diet until the ripening of the wild-rice and maize in the fall. "For these vermin," says Wilson, "the starlings search with great diligence in the ground, at the roots of plants in orchards and meadows, as well as among buds, leaves, and blossoms; and from their known voracity the multitudes of these insects which they destroy must be immense. Let me illustrate this fact by a short computation: If we suppose each bird on an average to devour fifty of these larvæ in a day (a very moderate allowance), a single pair in four months, the usual time such food is sought after, will devour upward of 12,000. It is believed that not less than a million pairs of these birds are distributed over the whole extent of the

United States in summer, whose food, being nearly the same, would swell the amount of vermin destroyed to 12,000,000,000. But the number of young birds may be fairly estimated at double that of their parents; and as these are constantly fed on larvæ for three weeks, making only the same allowance for them as for the older ones, their share would amount to 42,000,000,000, making a grand total of 54,000,000,000 of noxious insects destroyed in the space of four months by this single species! The combined ravages of such a hideous host of vermin would be sufficient to spread famine and desolation over a wide extent of the richest, best-cultivated country on the earth."

The yellow-headed blackbird, a kinsman of larger size, belongs properly north-west of Lake Superior, but frequently gets into Michigan and Illinois. The bright yellow head and neck make it very noticeable if seen. Its habits are essentially those of the redwing.

We have another set of blackbirds in the Atlantic States, of greater size than the *Agelæi*, commonly known as "crow" blackbirds, but called grakles in the books. There are several species, but none are greatly different from that too-common pest of our cornfields, the purple grakle.

The real home of the grakles, although along the edges of the swamps, is not among the reeds where the redwing and bobolink sit and swing, but rather in the bushes and trees skirting the muddy shores. They build their nests in a variety of positions, but usually a convenient fork in an alder-bush is chosen, twenty or thirty pairs often dwelling within a radius of a hundred feet. The nest is a rude, strong affair of sticks and coarse grass-stalks lined with finer grass, and looks very bulky and rough beside the neat structure of the redwing; which illustrates how much better a result can be produced by an artistic use of the same material. In the case of these, as well as the redwinged blackbirds, however, the female does not wear the jetty, iridescent coat which adorns the head of the family, and reflects the sunlight in a thousand prismatic tints, but hides herself and the home she cares for by affecting a dull, brown-black, streaked suit, assimilating her closely with the surrounding objects. This protective coloration of plumage is possessed by the females of many species of birds, which would be very conspicuous, and of course greatly liable to danger while incubating their eggs, if they wore the bright tints of the males. The tanager and indigo-bird afford prominent examples. Sometimes the crow blackbirds make their homes at a distance from the water, and occasionally they choose odd places, such as the tops of tall pine-trees, the spires of churches, martin-boxes in gardens, and holes in trees. The latter situation is one which the bronzed grakle of the Mississippi valley (var. *æneus*) especially makes use of.

Crow blackbirds' eggs are among the first on every boy's string, and until he gains experience the young collector supposes he has almost as many different species represented as he has specimens, so much do they differ, even in the same nestful, in respect to color, shape, and size. Their length averages about 1.25 by .90 of an inch, but some are long, slender, and pointed, while others are round, fat, and blunt at both ends. The ground color may be any shade of dirty white, light blue, greenish, or olive brown; the markings consist of sharply-defined spots and confused blotches, scratches, and straggling lines of obscure colors, from blue-black to lilac and rusty brown — sometimes scantily and prettily marbled upon the surface of the egg, and sometimes painted on so thick as to wholly conceal the ground color.

The crow blackbirds are in the advance-guard of the returning hosts of northward bound migrants, making their appearance in small scattering flocks, and announcing their presence by loud smacks frequently repeated. They obtain most of their food from the ground, and walk about with great liveliness, scratching up the leaves, turning over chips, and poking about the pastures for insects and seeds softened by the spring rains. Their destruction of insects—especially during May, when their young are in the nest—is enormous; yet their forays upon the cornfields, I fear, overbalance the good done the farmer by putting an end to grubs noxious to his crops.

"The depredations committed by these birds are almost wholly on Indian-corn at different stages. As soon as its blades appear above the ground after it has been planted, the grakles descend upon the fields, pull up the tender plant and devour the seeds, scattering the green blades around. It is of little use to attempt to drive them away with a gun: they only fly from one part of the field to another. And again, as soon as the tender corn has formed, these flocks, now. replenished by the young of the year, once more swarm in the cornfields, tear off the husks, and devour the tender grains." Wilson saw fields in which more than half the corn was thus ruined.

These birds winter in immense numbers in the lower parts of Virginia, North and South Carolina, and Georgia, sometimes forming one congregated multitude of several hundred thousand. On one occasion Wilson met, on the banks of the Roanoke, on the 20th of January, one of these prodigious armies of crow blackbirds. They arose, he states, from the surrounding fields with a noise like thunder, and, descending on the length of the road before him, they covered it and the fences completely with black: when they again rose, and after a few evolutions descended on the skirts of the high-timbered woods, they produced a most singular and striking effect. Whole trees, for a considerable extent, from the top to the lowest branches, seemed as if hung with mourning. Their notes and screaming, he adds, seemed all the while like the distant sounds of a great cataract, but in a musical cadence. This is a scene which may be paralleled every autumn in the grain districts of the West and South.

III.

WILD MICE.

When every stream in its pent-house
   Goes gurgling on its way,
And in his gallery the mouse
   Nibbleth the meadow hay;

Methinks the summer still is nigh,
   And lurketh underneath,
As that same meadow-mouse doth lie
   Snug in that last year's heath.
THOREAU.

WALKING about the fields, I come upon little pathways as plain as Indian trails, which lead in and out among the grass and weed-stalks, under Gothic arches the bending tops of the flowering grasses make, like roads for the tiny chariots of Queen Mab. These curious little paths branching here and there, and crossing one another in all directions, are the runways of the field-mice, along which they go, mostly after sunset, to visit one another or bring home their plunder; for the thieving little gray-coats of our cupboards, whose

58 *FRIENDS WORTH KNOWING.*

THE HOUSE-MOUSE.

bright eyes glance at us from behind the cheese-box, and who whisk away down some unthought-of hole, learned their naughty tricks from their many out-door cousins, whom we may forgive on the plea of their not knowing any better. Suppose I tell you about some of these same cousins who live in the woods and fields of the Northern half of the United States?

If you take the *o* and the *e* out of "mouse," you have left *mus*, which is the Latin word for mouse; but instead of saying "mousey," a Roman girl would have said *musculus*. Put the two together, and you have *Mus musculus*, the name we write when we want every person, whether he understands our language or not, to know that we mean the common house-mouse, for all the world is supposed to know something of Latin. This little plague was originally a native of some Eastern country, but has now spread all over the world, forgetting where he really does belong. Sometimes, in this country, he forsakes the houses and takes up a wild life in the woods.

Coming now to our true field-mice, there is first one which, to distinguish it from Old World kinds, is called in the books by Greek words which mean the "white-footed Western mouse" — *Hesperomys leucopus* — a very good name. A second sort is generally found in meadows through which brooks wander; and its Latin name, *Arvicola riparius*, just tells the whole story in two words; it is the "meadow-mouse." The third and last sort of wild mouse in Eastern America was first noticed near Hudson's Bay, and, being a great jumper, received the name of the "Hudsonian jumping-mouse"—*Jaculus hudsonius*.

These four mice differ in shape, color, size, and habits, and of the second and third there are several varieties in different parts of the country. The soft, brownish-gray coat of the house-mouse you know very well; or, if you do not, take the next one you catch and look at it closely. It is as clean as your pet squirrel, and just as pretty. See how dainty are the little feet, how keen the black beads of eyes, how sharp and white the fine small teeth, how delicate the pencillings of the fur!

Prettiest of all is the long-legged jumping-mouse. If you should look at a kangaroo through the wrong end of a telescope, you would have a very fair idea of our little friend's form, with hind-legs and feet very long and slender, and fore-legs very short; so that when he sits up they seem like little paws held before him in a coquettish way. His tail is often twice the length of his body, and is tipped with a brush of long hairs. He has a knowing look in his face, with its upright, furry ears and bright eyes. Being darkbrown above, yellowish-brown on the sides, and white underneath, with white stockings, he makes a gay figure among his more soberly dressed companions. Various names are given him; such as the deer-mouse, wood-mouse, jumping wood-mouse, and others.

The white-foot is somewhat larger than the house-mouse;

THE JUMPING MOUSE (JACULUS).

being about three inches long. It has a lithe, slender form, and quick movement; its eyes are large and prominent, its nose sharp, and its ears high, round, and thin. The fore-feet are hardly half as long as the hinder ones, and the tail is as long as, or longer than, the body, and covered with close hairs. The fur is soft, dense, and glossy, reddish-brown above and white below, while the feet are all white.

The most ill-looking of the lot is the meadow-mouse, which reminds me of a miniature bear. Its coat is dirty brownish-black, not even turning white in winter; its head is short, and its nose blunt; all its four feet are short, and its tail is a mere stump, scarcely long enough to reach the ground. Nevertheless, it is a very interesting mouse, and able to make an immense deal of trouble.

In general habits the three wild ones are pretty much alike, though some prefer dry, while others choose wet, ground; some keep chiefly in the woods, others on the prairies, and so on. All the species burrow more or less, and some build elaborate nests. Their voices are fine, low, and squeaking, but the meadow-mouse is a great chatterbox, and the white-foot has been known more than once really to sing tunes of his own very nicely. Each one manifests immense courage in defending its young against harm; but I believe only the meadow-mice are accused of being really ferocious, and of waging battles constantly among themselves. Their food is the tender stems of young grasses and herbs, seeds, nuts, roots, and bark, and they lay up stores of food for the winter, since none become torpid at that season, as is the habit of the woodchuck and chipmonk, except the jumping-mouse. This fellow, during cold weather, curls up in his soft grass blankets underground, wraps his long tail tightly about him, and becomes dead to all outward things until the warmth of spring revives him, which is certainly an easy and economical way to get through the winter! They also eat insects, old and young, particularly such kinds as are hatched underground or in the loose wood of rotten stumps; but their main subsistence is seeds and bark, in getting which they do a vast deal of damage to plants and young fruit-trees with those sharp front teeth of theirs.

The field-mice make snug beds in old stumps, under logs, inside stacks of corn, and bundles of straw; dig out galleries below the grass roots; occupy the abandoned nests of birds and the holes made by other animals; and even weave nests of their own in weeds and bushes. They live well in captivity, and you can easily see them at work if you supply materials.

In tearing down old buildings the carpenters often find

between the walls a lot of pieces of paper, bits of cloth, sticks, fur, and such stuff, forming a great bale, and know that it was once the home of a house-mouse. You have heard various anecdotes of how a shopkeeper misses small pieces of money from his till, and suspects his clerk of taking it; how the clerk is a poor boy who is supporting a widowed mother, or a sister at school, and the kind-hearted shopkeeper shuts his eyes to his suspicions, and waits for more and more proof before being convinced that his young clerk is the thief; but, as the money keeps disappearing, at last he must accuse the clerk of taking it. Then the story tells how, in spite of the boy's vehement and tearful denial, a policeman is called in to arrest him, and when everything has been searched to no purpose, and he is about being taken to the police-station, how, away back in a corner is discovered a mouse's nest made of stolen pieces of ragged currency—ten, twenty-five, and fifty-cent pieces. Then everybody is happy again, and the story ends with a capital moral!

More than one such stolen house the mice have really built, and sometimes their work has destroyed half a hundred dollars, and caused no end of heartaches. Their little teeth are not to be despised, I assure you. I believe one of the most disastrous of those great floods which in past years have swept over the fertile plains of Holland was caused by mice digging through the thick banks of earth, called dikes, which had been piled up to keep the sea back. In this case, of course, the mice lost their lives by their misdeeds, as well as the people, sharing in the general catastrophe. They hardly intended this; but

> "The best-laid plans o' mice and men
> Gang aft agley."

It was by the gnawing of a ridiculous little mouse, you remember, that the lion in the fable got free from the net in which the king of beasts found himself caught.

Sometimes the house-mouse goes out of doors to live, and forgets his civilization; while, on the other hand, the woodland species occasionally come in doors and grow tame. At the fur-trading posts about Hudson's Bay wild mice live in the traders' houses; and Thoreau—the poet, naturalist, and philosopher, whom all the animals seemed at once to recognize as their friend—wrote this beautiful story of how a white-footed mouse made friends with him when he lived all alone in the woods by Walden Pond, near Concord, Massachusetts:

"The mice which haunted my house were not the common ones, which are said to have been introduced into the

THE WHITE-FOOTED MOUSE.

country, but a wild native kind not found in the village. I sent one to a distinguished naturalist, and it interested him much. When I was building, one of these had its nest underneath the house, and before I had laid the second floor and swept out the shavings, would come out regularly at lunch-time and pick up the crumbs at my feet. It probably had never seen a man before; and it soon became quite familiar, and would run over my shoes and up my clothes. It could readily ascend the sides of the room by short impulses, like a squirrel, which it resembled in its motions. At length, as I leaned my elbow on the bench one day, it ran up my clothes and along my sleeve, and around and around the table which held my dinner, while I kept the latter close, and dodged and played at bo-peep with it; and when at last I held still a piece of cheese between my thumb and finger, it came and nibbled it, sitting in my hand, and afterward cleaned its face and paws like a fly, and walked away."

Mice are full of such curiosity. They poke their noses into all sorts of places where there is a prospect of something to eat, and sometimes, failing to find so good a friend as Mr. Thoreau, meet the fate which ought to be the end of all poking of noses into other people's affairs—they get caught. I remember one such case which Mr. Frank

Buckland has related. When oysters are left out of water for any length of time, especially in hot weather, they always open their shells a little way, probably seeking a drink of water. A mouse hunting about for food found such an oyster in the larder, and put his head in to nibble at the oyster's beard; instantly the bivalve shut his shells,

THE MOUSE AND THE OYSTER.

and held them together so tightly by his strong muscles that the poor mouse could not pull his head out, and so died of suffocation. Other similar cases have been known.

The most common of all our field-mice is the short-tailed meadow-mouse, the *Arvicola*. I find it in the woods, out on the prairies, and in the hay-fields. In summer these little creatures inhabit the low, wet meadows in great numbers. When the heavy rains of autumn drive them out, they move to higher and drier ground, and look for some hillock, or old ant-hill, under which to dig their home. In digging they scratch rapidly with the fore-feet a few times, and then throw back the earth to a great distance with the hind-feet, frequently loosening the dirt with their teeth, and pushing it aside with their noses. As the hole grows deeper (horizontally) they will lie on their backs and dig overhead, every little while backing slowly out and shoving the loose earth to the entrance. These winter burrows are only five or six inches below the surface, and sometimes are simply hollowed out under a great stone, but are remarkable for the numerous and complicated chambers and side passages of which they are composed. In one of the largest rooms of this subterranean house is placed their winter bed, formed of fine dry grasses. Its shape and size are about that of a foot-ball, with only a small cavity in the centre, entered through a hole in the side, and they creep in as do Arctic travellers into their fur-bags.

> "Thou saw the fields laid bare an' waste,
> An' weary winter comin' fast,
> An' cosy here, beneath the blast
>   Thou thought to dwell."

Here five or six young mice are born, and stay until the coming of warm weather, by which time they are grown, and go out to take care of themselves. Sometimes one of them, instead of hunting up a wife and getting a home of his own, will wander off by himself and live alone like a hermit, growing crosser as he grows older.

In the deepest part of the burrow is placed their store of provisions. Uncover one of these little granaries in November, before the owners have used much of it, and you might find five or six quarts of seeds, roots, and small nuts. Out on the prairie this store would consist chiefly of the round tubers—like very small potatoes—of the spike-flower, a few juicy roots of some other weeds and grasses, bulbs of the wild onion, and so forth. If a wheat or rye patch was near, there would be quantities of grain; and if you should open a nest under a log or stump in the woods, you might discover a hundred or so chestnuts, beech-nuts, and acorns, nicely shelled. All these stores are carried to the burrows, often from long distances, in their baggy cheeks, which are a mouse's pockets, and they work with immense industry, knowing just when to gather this and that kind of food for the winter. A friend of mine, who had a farm near the Hudson River, had a nice field of rye, which he was only waiting a day or two longer to harvest until it should be quite ready. But the very night before he went to cut it, the mice stole a large portion of the grain and carried it off to their nests in the neighboring woods. Hunting up these nests he got back from two of them about half a bushel of rye, which was perfectly good. Sometimes they build nests in the russet corn-shocks left standing in the sere October fields, and store up there heaps of food, although there may be no necessity, so firmly fixed in their minds is the idea of preparing for the future. But they eat a great deal, and their stores are none too large to outlast the long, dreary months, when the ground is frozen hard, and the meadows are swept by the wintry winds, or packed under a blanket of snow.

The English field-mouse, which is very much like our own, has "a sweet tooth," and searches for the nests of the bumblebees in order to get the comb and honey.

The *Arvicola* and *Jaculus* seem to be the greatest diggers, while the *Hesperomys* prefers a home above-ground, and constructs its dwelling much like the squirrel's. Sometimes it takes up its abode in deserted birds'-nests, such as those of the cat-bird, red-winged blackbird, wood-thrush, and red-eyed vireo. A cradle-nest of the last-named bird, which had been thus used by a white-footed mouse, was found toward the end of August, 1875, on the border of a thick forest on the Blue Ridge, by Mr. Spencer Trotter, of Philadelphia. This nest, which—second tenant and all—is given as the frontispiece, hung from the extremity of a young tree a few feet from the ground; and the mouse had completely filled the inside with dry grass, leaving only enough room to squeeze into a comfortable bed in the bottom. The mouse was asleep when found, as is its habit in the daytime, and moved away rather sluggishly.

Not long ago, I received a pleasant letter from Mr. John Burroughs, in which he said: "The other day I found the nest of the white-footed mouse. Going through the woods, I paused by a red cedar, the top of which had been broken off and lopped over till it touched the ground. It was dry, and formed a very dense mass. I touched a match to it to see it burn, when, just as the flames were creeping up into it, out jumped or tumbled two white-footed mice, and made off in opposite directions. I was just in time to see the nest before the flames caught it—a mass of fine dry grass, about five feet from the ground, in the thickest part of the cedar top." This was in the Catskills.

From their tunnels, nests, and granaries, innumerable runways, such as I spoke of before, traverse the neighborhood, crossing those from other burrows, and forming a complete net-work all over the region. The mice do not flock together like the prairie dogs, but, where food is plenty, many nests will often be found closely adjacent. They are sociable little folk, and no doubt enjoy visiting and gossiping with one another. The little paths are their

LEAVING HOME.

roadways from one burrow to another, and from the place where the tenderest grasses grow to their storehouses. These tiny roads are formed by gnawing clean away the grass stubble, and treading the earth down smooth; while the heads of the grasses arching over on each side conceal the scampering travellers from the prying eyes of owls, hawks, and butcher-birds, ever on the watch for them. The mice seem fully to understand their danger, cautiously going under a tuft of grass or a large leaf instead of over it, and avoiding bare places. In winter their paths are tunnelled under the snow, so that they are out of sight; and they always have several means of escape from their burrows. You know the old song says,

> "The mouse that always trusts to one poor hole,
> Can never be a mouse of any soul."

A trotting, gliding motion is the gait of the *Arvicola*, but the white-foot gallops along, jumping small objects, and leaping from one hillock to another, while the kangaroo-mouse springs off his hind-feet, and progresses in a series of long leaps, which carry him over the ground like a race-horse.

But the life of one of our favorites is not all frisking about under the fragrant flowers, or digging channels through shining sand and crystal snow. He has his labor and trials and trouble like the rest of us. If "a man mun be eäther a man or a mouse," it would be hard choosing between them, so far as an easy time is concerned! The

THE FIGHT WITH THE SNAKE.

gathering of his food, and the building of his house, costs him "mony a weary nibble," and he must constantly be on the alert, for dangers haunt him on every side. One of his enemies is the snake, all the larger sorts of which pounce upon him in the grass, lie in wait for him in his highway, or steal into his burrow and seize his helpless young, in spite of the frantic fighting of the father, and the stout attempts of the mother to drag her little ones away into safety. A gentleman in Illinois once saw a garter-snake pass rapidly by with a young meadow-mouse in its mouth. Presently an old meadow-mouse came out of the tall grass in pursuit of the snake, which she finally overtook and instantly attacked. The snake stopped, disgorged its prey, and defended itself by striking at its assailant, which appeared to be beating it, when both animals were killed by the gentleman watching. I am sorry the incident ended so tragically. The courage and affection of the little mother deserved a better reward, and even the garter-snake was entitled to some sympathy.

Probably our snakes depend more upon catching mice than upon any other resource for their daily food, and they hunt for them incessantly. Most of the mice have the bad habit of being abroad mainly at night; so have the snakes; and the mice thus encounter more foes, and fall an easier prey, than if they deferred their ramblings until daylight. Being out nights is a bad practice! The prairie rattlesnakes are especially fond of mice; minks, weasels, skunks, and badgers eat as many as they can catch, and this probably is not a few; domestic cats hunt them eagerly, seeming to prefer them to house-mice—no doubt they are more sweet and delicate; foxes also enjoy them; dogs and wolves dig them out of their burrows and devour them; prairie fires burn multitudes of them, and farmer-boys trap them. But, after all, perhaps their chief foes are the flesh-eating birds. I hardly ever take a walk without finding the remains of an owl's or hawk's dinner where our little subject has been the main dish.

We have in this country two black, white, and gray birds called shrikes, or butcher-birds, which are only about the size of robins, but are very strong, brave, and noble in appearance. These shrikes have the curious habit of killing more game than they need, and hanging it up on thorns, or lodging it in a crack in the fence or the crotch of a tree. They seem to hunt just for the fun of it, and kill for the sake of killing. Now their chief game is the unhappy field-mouse; and in Illinois they are known as "mouse-birds." They never seem to eat much of the flesh of their victims, generally only pecking their brains out, but murder an enormous number, and keep up the slaughter through the whole year; for when the loggerhead shrike retreats southward in the autumn, the great northern shrike comes from British America to supply his place through the winter. Then all the hawks, from the nimble little sharp-shinned to the great swooping buzzard, prey upon mice, and in winter hover day after day over the knolls where they have been driven by floods in the surrounding lowlands, pouncing upon every one that is imprudent enough to show his black eyes above ground. As for the marsh-hawk, it regularly quarters the low fields like a harrier, and eats little but mice. The owls, too, are constantly after them, hunting them day and night, on the prairies and in the woods, esteeming them fine food for the four owlets in the hollow tree hard by; while the sand-hill crane and some of the herons make a regular business of seeking the underground homes, and digging out the timorous fugitives with their pick-axe beaks. In addition to all the rest, the farmer everywhere persecutes the mouse, as a pest to his orchards and crops.

Has the poor little animal, then, no friends whatever? Very few, except his own endurance and cunning; yet he is already so numerous, and increases so rapidly, that all his enemies have not been able to rid the earth of him, but only to keep him in check, and thus preserve that nice balance of nature in which consists the welfare of all.

An important part of the history of these pretty wild mice would be untold if I were to say nothing about the mischief they do to the farmer's fields and fruit-trees. From the story I have related of the little "thieves in the night" who stole my friend's rye, and of their underground stores, you may guess how they make the grain-fields suffer. It is done so quietly and adroitly, too, that few are ever caught at it, and much of the blame is put on the moles, squirrels, and woodchucks that have enough sins of their own to answer for. The meadow-mouse of Europe, which is very like our own, forty or fifty years ago came near causing a famine in parts of England, ruining the crops before they could get fairly started, and killing almost all the young trees in the orchards and woods. More than 30,000 of the little rascals were trapped in one month in a single piece of forest, besides all those killed by animals. About 1875, again, a similar disaster was threatened in Scotland, where millions of mice appeared, and gnawed off the young grass at the root just when it should have been in prime condition for the sheep; and when that was all gone they attacked the garden vegetables. The people lost vast numbers of sheep and lambs from starvation, and thousands of dollars' worth of growing food; but, finally, by all together waging war upon them, the pests were partially killed off. The mice did not in either case come suddenly, but had been increasing steadily for years previous, because the game-keepers had killed so many of the "vermin" (as owls, hawks, weasels, snakes, etc., are wrongly called), which are the natural enemies of the mice, and keep their numbers down. Farmers are slow to learn that it doesn't pay to kill the birds or rob their nests; but the boys and girls ought to understand this truth and remember it. In this country the greatest mischief done by the field-mice is the gnawing of bark from the fruit-trees, so that in some of the Western States this is the most serious difficulty the orchardist has to contend with. Whole rows of young trees in nurseries are stripped of their bark, and of course die; and where apple-seeds are planted, the mice are sure to dig half of them up to eat the kernels. This mischief is mainly done in the winter, when the trees are packed away from the frost; or if they are growing, because then the mice can move about concealed under the snow, and nibble all the bark away up to the surface. Rabbits get much of the credit of this naughty work, for they do a good deal of it on their own account. The gardener has the same trouble, often finding, when he uncovers a rare and costly plant in the spring, that the mice have enjoyed good winter-quarters in his straw covering, and have been gnawing to death his choice roses. Millions of dollars, perhaps, would not pay for all the damage these small creatures thus accomplish each year in the United States, and I fear they will become more and more of a plague if we continue to kill off the harmless hawks, owls, butcher-birds, and snakes, which are the policemen appointed by Nature to look after the mice, and protect us against them.

In captivity the wild mice, especially the white-footed *Hesperomys*, make very pretty pets; and one can easily study all their ways by giving them earth in which to burrow, and the various sorts of food in which they delight.

IV.

AN ORNITHOLOGICAL LECTURE.

I HAD almost written my title, unconsciously, Beautiful Birds, for they have become symbols to us of all that is blithesome and free. No one of all the classes of animals is more worthy of attention, or more easily studied. Including within their number every variety of costume and shape; present everywhere, and at all times; making us their confidants by coming to our door-steps, or awaiting us with newer and newer surprises if we go to the remote woods, the pathless ocean, or snowy mountain; marshalling their ranks over our heads, coming and going with the seasons, and defying our pursuit; surely, here is something for the poet and artist, as well as the naturalist, to think upon.

But a bird is something more than a flitting fairy, or an incarnation of song. It has substance and form; it moves swiftly, mysteriously from place to place, and looks out carefully for its own protection and subsistence; it cunningly builds a home, where it raises its young and teaches

HAUNT OF THE HERON.

them to care for themselves. The how and why of some of these incidents of bird-life I want to tell you—I say some, for after all many of the ways of our familiar birds are unexplained.

The most prominent fact about a bird is a faculty in

which it differs from every other creature except the bat and insects — its power of flying. For this purpose, the bird's arm ends in only one long slender finger, instead of a full hand. To this are attached the quills and small feathers (coverts) on the upper side, which make up the wing. Observe how light all this is: in the first place, the bones are hollow; then the shafts of the feathers are hollow; and, finally, the feathers themselves are made of the most delicate filaments, interlocking and clinging to one another with little grasping hooks of microscopic fineness.

THE KINGFISHER.

Well, how does a bird fly? It seems simple enough to describe, and yet it is a problem that the wisest in such matters have not yet worked out to everybody's satisfaction. This explanation, by the Duke of Argyle, appears to me to be the best: An open wing forms a hollow on its underside like an inverted saucer; when the wing is forced down, the upward pressure of the air caught under this concavity lifts the bird up, much as you hoist yourself up between the parallel bars in a gymnasium. But he could never in this way get ahead, and the hardest question is still to be answered. Now, the front edge of the wing, formed of the bones and muscles of the forearm, is rigid and unyielding, while the hinder margin is merely the soft flexible ends of the feathers; so, when the wing is forced down, the air under it, finding this margin yielding the easier, would rush out here, and, in so doing, would bend up the ends of the quills, pushing them forward out of the way, which, of course, would tend to shove the bird ahead. This process, quickly repeated, results in the phenomenon of flight.

The vigor and endurance that birds upon the wing display is astonishing. Nearly all the migratory species of Europe must cross the Mediterranean without resting. Many take the direct course between the coast of Africa

AN ORNITHOLOGICAL LECTURE. 89

and England, which is still farther. Our little bluebird pays an annual visit to the Bermudas, six hundred miles from the continent, and Wilson estimated its apparently

SUMMER YELLOW-BIRDS.

very moderate flight at much more than a mile a minute. Remarkable stories are told of the long flights tame falcons have been known to take—one going a thousand three hundred miles in a day. Yarrell mentions carrier-pigeons that flew from Rouen to Ghent, one hundred and fifty miles, in an hour and a half; but this speed is surpassed by our own wild pigeons, which have been shot in New York before the rice they had picked in Georgia had been digested. It is ascertained that a certain warbler must wing its way from Egypt to Heligoland, one thousand two hundred miles, in one night, and it is probable that martins endure equal exertion every long summer's day, in their ceaseless pursuit of insects. Taking, then, one hundred miles per hour as the rate of flight during migrations, we need not be surprised that representatives of more than thirty species of our wood-birds have been shot in the British Isles, since they could well sustain the sixteen hundred miles between Newfoundland and Ireland.

"A good ornithologist," says White of Selborne, "should be able to distinguish birds by their air, as well as their colors and shape, on the ground as well as on the wing, and in the bush as well as in the hand." Almost every family of birds has its peculiarities of manner. Thus, the kites and buzzards glide round in circles with wings expanded and motionless; marsh-hawks or harriers fly low over meadows and stubble-fields, beating the ground regularly. Crows and jays lumber along as though it were hard work; and herons are still more clumsy, having their long necks and longer legs to encumber them. The woodpecker's progress is in a series of long undulations, opening and closing the wings at every stroke. Our thistle-loving goldfinch also flies this way, but the most of the *Fringillidæ* (finches, sparrows, etc.) have a short, jerking flight, accompanied with many bobbings and flirtings. Warblers and fly-catchers fly high up, smoothly and swiftly. Swallows and night-hawks seem to be mowing the air with cimitar wings, and move with surprising energy. On the ground, most small birds are hoppers, like the sparrows, but a few, like the robin and water-thrush, truly and gracefully walk, and the "shore-birds" are emphatically runners. Among all sorts, queer movements are assumed in the love season, not noticeable at other times.

There is no part of the world where the feathered tribe is not represented; but no two quarters of the globe, and scarcely any two places a hundred miles apart, have precisely the same sort of birds, or in similar abundance. There are several reasons for this: first, the influence of climate. Birds provided with the means of resisting the extreme cold of northern regions would be very uncomfortable under a southern sun. The geographical distribution of plants has long been recognized, but it is only recently that a like distribution of birds has been proved

YELLOW-BREASTED CHATS.

to exist. Moreover, oceans and high mountain chains limit the range of many kinds. Europe and America have scarcely any species in common, save of water-birds and large hawks. Those from the Pacific coast are essentially different from those found in the Mississippi Valley. Each district has a set of birds— and other animals as well—peculiar to its peculiar geography. Another great circumstance, determining the presence or absence of certain birds in the breeding season, is the abundance or scarcity of suitable food, not only for themselves, but also for their young; as the food of birds at that time is often very different from their ordinary diet, it requires a close acquaintance with them to prophesy confidently what birds would be likely to be found breeding at a given point.

But few birds remain in the same region all the year round. Out of about two hundred and seventy-five species occurring in New England or New York in June, only twenty-five or so stay throughout the year; of these forty or fifty come to us in winter only, leaving us two hundred and twenty-five species of spring birds, half of which number merely pass through to their northern breeding-places. With this disparity, no wonder that we look for the return of the birds, and hail with delight the bluebird calling to us through clear March mornings, the velvet-coated robins, the battalions of soldierly cedar-birds, the ghostly turtle-doves sighing their surging refrain, the pewees, and thrushes, and golden orioles, till at last, amid the bursting foliage and quickness of May life, a full host of brilliant choristers holds jubilee in the sunny tree-tops.

In a very few days, as suddenly and mysteriously as they came, half the gay company has passed us, going farther north to breed. Could we follow this army, we should find it thinning gradually, as one species after another found its appropriate station — a part in upper New England and Canada, many about Hudson's Bay; while not a few (water-birds especially) would lead us to the very shores of Arctic fjords. For them the summer is so short that ice and snow start them south before we have any thought of cold weather. On their way they pick up all the Labrador and Canada birds, re-enforced by their young, so that an even greater army invades our woods amid the splendor of October than made them ring in the exuberance of June. Then our own birds catch the infection, and singly, or in squads, companies, and regiments, join the great march to the savannas of the Gulf States, the table-lands of Central America, and on even to the jungles of the Orinoco. What a wonderful perception is that which teaches them to migrate; tells them just the day to set out, the proper course to take, and keeps them true to it over ocean and prairie, and monotonous forests, and often in the night! That the young, learning the route from the parent, remember it, would be no less remarkable were it true, which it probably is not; for many species seem to go north by one route, as along the coast, and return by another west of the Alleghanies, or *vice versa*. In proceeding northward, the males go ahead of the females a week or so; returning in the fall, the males again take the lead, and the young bring up the rear. Yet

A JUNE MORNING.

there are many exceptions to this rule, for with not a few birds the males and females travel together; and with some, old and fully plumaged males are the last to arrive. All birds migrate more or less—even such, like the crow and song-sparrow, as stay with us through the year; for we probably do not see the same individuals both winter and summer. Even tropical birds move a little way from the equator, and back again with the season; and in mountainous regions most of the birds, and many small quadrupeds,

THE HUMMING-BIRD'S NEST.

have a vertical migration only, descending to the valleys in winter, and reascending to the summits in summer—difference of altitude accomplishing the same climatic results as a change in latitude.

We can see various causes of these migrations, some of

"ONLY A CAT-BIRD."

which have already been suggested, but the chief cause seems to be the necessity of their accustomed food. We find that those birds which make the longest and most complete migrations are insect and honey-eaters; while the graminivorous and omnivorous birds, and such, like the titmouse and nut-hatch, as subsist on the young of insects to be found under the bark of trees, go but a short distance to escape inclement weather, or do not migrate at all. Sportsmen recognize the fact that the snipe and woodcock have returned, not because the rigorous winter days are wholly passed, but because the frost is sufficiently out of the ground to allow the worms to come to the surface; and know that in warm, springy meadows these birds may often be found all through the year. Man no doubt influences the migratory habits of birds. To many he offers inducements in the shelter, and in the abundance of insects which his industry occasions, to linger later in the fall than was their wont, and return earlier in the spring. While, on the contrary, the persecution which the shy wild-fowl have received has caused them generally to repair to secluded breeding-places, far north of their haunts of fifty years ago. But the migrations of most birds are somewhat irregular, and we have so few reliable data that we can hardly yet determine the laws which govern their seasonal movements, much less assert the ancient origin of the "migratory instinct," so called, or state the varied influences that have led to the present powerful habit, and have pointed out the routes which the flocks now follow, spring and fall. The geologist must aid the zoological student in solving these problems.

The true home of a bird, then, is where it rears its young, even though it be not there more than a third of the year,

EAGLES.

and everywhere else it is merely a traveller or *migrant*. Should you then, after say two years of observation, want to write down a list of the birds inhabiting your district— and you would thus be doing a real service to science—it is important that you mention whether each bird breeds there, passes through spring and autumn, or is only a winter visitor.

Perhaps there is no animal in the world that comes nearer to man's heart, and seems more akin, than the bird, be-

THE PLOVER.

cause of its beautiful home-life, and the loving care with which it anticipates and provides for its brood. There is a charm about the nest of a bird that does not linger about the hive of the wild bees, the burrow of the woodchuck, or the dome of the musk-rat. It is more a *home* than any of them. The situation varies as much as the birds themselves. Trees, however, form the most common support: among the tip-top branches of them warblers fix their tiny cradles; to the outer drooping twigs of them orioles and vireos can swing their hammocks; upon their stout horizontal limbs the thrushes and tanagers may come and build; against the trunk, and in the great forks, hawks, and crows, and jays will pile their rude structures; and in the cracks and crannies, titmice, nut-hatches, and woodpeckers clean out old holes, or chisel new, in which to deposit their eggs. But most of the large birds of prey inhabit lone crags, making an eyrie which they repair from year to year for the new brood. The ground, too, bears the less pretentious houses of sparrows and larks, and the scattered eggs of sand-pipers, gulls, and terns; the marshes are occupied by rails, herons, and ducks; the banks of rivers are burrowed into by kingfishers and sand-martins; so that almost every conceivable position is adopted by some bird or another, and its peculiar custom usually, though not by any means invariably, adhered to by that species. A curious instance of change in this respect is shown by the two barn-swallows and the chimney-swallow, which, before the civilization of this country, plastered their nests in caves, and in the inside of hollow trees, as indeed they yet do in the far north-west. In the materials used, and the construction of the nest, birds adapt themselves largely to circumstances. In the Northern States, for example, the Baltimore oriole uses hempen fibres, cotton twine, *et cætera*, for its nest; but in the heat of Louisiana the same pouch-shaped

THE WOOD-PEWEE.

structure is woven of Spanish moss, and is light and cool. The intelligence and foresight that some birds exhibit in their architecture prove reason rather than instinct, as we popularly use these words; while others are so stupid as to upset all our respect for their faculties of calculation. Both sexes usually help in building the nest, and work industriously at it till it is ready for the eggs—sometimes finishing it even after the female has begun to sit.

The best known birds probably are such famous songsters

TURKEY-BUZZARDS.

as the nightingale and the skylark; and because these and our canaries are foreign, most persons suppose that we have no equally fine songsters of our own. Let a doubter go into the June woods only once! June is harvest-month for the ornithologist. Then the birds are dressed in their best, are showing off all their good points to their lady-loves, are building their nests, and—being very happy—are in full song. Morning and evening there is such a chorus as makes the jubilant air fairly quiver with melody, while all day you catch the *yeap* of pigmies in the tree-tops, the chattering and twittering of garrulous sparrows and swallows, and the tintinnabulation of wood-thrushes. I cannot even name all these glorious singers. Perhaps the many-tongued mocking-bird stands at the head of the list; possibly the hermit-thrush, whose song is of "serene religious beatitude," or the blue grossbeak or winter wren. As you choose. The bird you think pre-eminent to-day will be excelled to-morrow, and you will refuse to distinguish between them for the love and admiration you bear them all.

V.

OUR WINTER BIRDS.

Not often in the genial days of early and late summer, or even in the torrid heat of its middle months, do we recall winter with pleasure, or wish ourselves surrounded by its scenes; while, on the contrary, the dark hours of the long winter evenings are often enlivened with reminiscences of balmy weather, the fireplace is adorned with bouquets of dried flowers, and every indication of returning spring is eagerly welcomed. Nothing is more precious to the eye, weary of the desolation which snow and ice bring to the landscape, than the winter birds, whose bright forms alone diversify the bare and colorless world, and whose cheery notes alone break the stillness and apparent immobility of Nature. They always carry a bit of the June sunshine about with them, and dropping it from their wings, like seed, wherever they flit, seem thus to preserve the season through the ravages of winter, to which all else succumbs. Some words about them may, therefore, help to keep the sense of summer alive in our hearts through this midnight of the year.

Most persons are surprised when told of the large number of these feathered friends which begin the new year with us; for in January, in the near neighborhood of New York city, over fifty species appear with more or less regularity. They comprise two classes: those which reside in our fields the year round, like the bluejay; and such, like the snow-flake, as are driven to our milder climate by the severity of a Northern winter that even their arctic-bred, hardy constitutions are unable to endure. The members of the latter class visit us in varying numbers, but are especially numerous in snowy seasons.

It is probably less a fear of the dreadful temperature, even in the frigid zones, which compels the birds to seek our milder latitudes, than the inability to obtain food when snow buries the seed-bearing weeds and sends the smaller animals to their hibernacula, and the increasing darkness of the long arctic night shuts out from view what the snow has not covered. All birds—or almost all—on their southward migration, fly at night, resting during the day. We have the most abundant evidence of this; and it has occurred to me that possibly it is the deepening darkness of high latitudes which first warns them off; that the natural recurrence of night seems to them like being overtaken by the darkness which they thought they had left behind, but which they must again flee; that, therefore, they keep upon the wing until each morning's light, supposing that they have thus again and again outstripped the pursuing gloom, until they reach a region of abundant food, and perhaps learn wisdom from its resident birds. I will confess that I do not myself put much faith in this theory, but a curious and sustaining fact is, that the northward migration, in spring, is mostly accomplished by day-journeys instead of at night.

Whatever the motive, no sooner has the crowd of autumnal migrants, with rustling wings and faint voices, swept through our woods—slowly during the long, mellow October days, when the earth seems to stand still, and the seasons to be in equipoise; swiftly when the first blast of November sends them skurrying onward with the deadened leaves—than their places are taken by the brave little fellows whose fame I celebrate.

Taking my way to the woods some bright, still morning in January, when the snow is crisp and the ice in the swamps firm, I shall find the sombre fields full of a life of their own well worth my while to see, even if the exhilaration of the walk does not prove reward enough. Here on this fence-rail is the track of a squirrel, and in the corner of the rail and rider is the half-eaten body of a chickadee which some butcher-bird has hung up. How the dry wood creaks as I climb over, and how resonant is that dead ash under the vigorous hammer of the little woodpecker

SNOW-BUNTING.

whose red crest glows like a spark of fire against the white limb! Around this spice-bush the mice have been at work, nibbling the bark off up to the surface of the snow, and we can see the entrance to their tunnel. This path, trod bare by the cows, leads to the hilly brush-pasture where the southern sun shines all the afternoon, and thither let me follow.

Sunny hill-sides, the wooded banks of creeks, the hedge-rows and brier-grown fences along the country roads, are all favorite places for the winter birds. Here come the sparrows and finches, the winter wren and rare cardinal, skulking about the thickets, hopping through the dead fern-brakes, threading the mazy passages of the log-heaps and brush-piles ready to be burnt in the spring, coming out upon the fence-post or way-side trees to sing their morning roundelay and take their daily airing in pleasant weather. In the open meadows are the grass-finches, snow-birds, and the few robins and medlarks that stay with us; in the edge of the woods the bluejay, flicker, and butcher-bird; in the orchards and evergreens the crossbills, the pine grossbeaks, red-polls, and cedar-bird; the deep woods shelter the tiny nuthatches, titmice, and the little wood-peckers; the open sky affords space for the birds of prey, and the sea-shore harbors for the gulls, sea-ducks, and fish-hawk. Such are the chosen resorts of the different varieties, yet, of course, we shall occasionally meet all everywhere, and sometimes spots apparently most favorable will be totally uninhabited. In very severe weather the wildest birds are often compelled to come close to the house and barn in search of out-door relief from gentle hands.

"How do the birds manage at night and in tempestuous weather?" is a question often asked me.

The time is not long passed when it was universally believed that many of them hibernated—especially the swallows—burying themselves in the mud like frogs, or curling up in holes in rocks like the bats; and the common phenomenon of the appearance of a few summer birds during "warm spells" in winter was assumed to prove that they had been torpid, but had waked up under the genial warmth, as bats often do. It was not three months ago that I saw in an English newspaper a letter from a man who claimed to have found a hedge-sparrow (I think) torpid somewhere in the mud. But the search for proofs of this theory discovered that the birds supposed to hibernate migrated, while of the birds which remained in this latitude through the cold months we saw more in warm, fine weather, for the natural reason that then they forsook the sheltered hollows and cosy recesses of the woods where they had retreated during stormy days, and came out into the sunlight. Dense cedars and the close branches of small spruces and other evergreens afford them good shelter, and thickets of brambles are made use of when these are not to be found; hollow trees are natural houses in which large numbers huddle, and the cave-like holes under the roots of trees growing on steep banks become favorite hospices. The grouse plunges through the snow down to the ground, where it scrapes a "form," or crawls under the hemlock and spruce boughs that droop to the earth with the weight of snow, and allows the white mantle to drift over it, subsisting the while on the spruce-buds; when the storm ceases it can easily dig its way out, but sometimes a rain and hard frost follow, which make such a crust on the snow that it cannot break up through, and so it starves to death. The more domestic sparrows, robins, and flickers burrow into the hay-mow, find a warm roost in the barn near the cattle, or, attracted by the warmth of the furnace, creep under the eaves or into a chink next the chimney of the greenhouse or country dwelling. The meadow-lark and quail seek out sunny nooks in the fields and crouch down out of the blast; while the woodcock hides among the moss and ferns of damp woods where only the very severest cold can chain the springs. Along the coast many birds go from the interior to the sea-shore in search of a milder climate.

Nevertheless, in spite of all these resources in the way of shelter; in spite of their high degree of warmth and vitality, probably not exceeded by any other animal; in spite of the fact that they can draw themselves up into a perfect ball of feathers which are the best of clothing, and that they can shelter themselves from the driving storm, it appears that birds often perish from cold in large numbers. Ordinarily, birds seem able to foretell a change of weather, and prepare. The reports of the United States weather bureau certainly show, that, during the fall and winter, the ducks, geese, cranes, crows, and other notable species—and apparently generally—abandon their former haunts upon the approach of a cold wave or hard winter storm for more southern localities, often passing beyond the reach of the severity of such storms, though taking their departure only a few hours before these unfavorable changes. Resident species, not caring, or not able, to run away to warmer latitudes, ought to know enough to hide away from the fury of the gale; and they do. But sometimes there come sudden, unpresaged changes—cold, icy gales, which charge down upon us after thawing-days, converting the air, which was almost persuading the grass to revive, into an atmosphere that cuts the skin like the impinging of innumerable particles of frost, and shrivels every object with cold, or buries it under dry and drifting snow. Then it is that the small birds, caught unprepared, suffer. At first, such as are overcome seem unusually active, running about apparently in search of food, but taking little notice of one's approach. "Should one attempt to fly," writes a recent observer, "it immediately falls on its back as if shot. The legs and toes are stretched out to their

BROWN CREEPER.

farthest extent, and are quite rigid; the eyes protrude, are insensible to the touch, and the whole body quivers slightly. It remains in this state from one to two minutes, when it recovers suddenly, and seems as active as before. If taken in the hand, it will immediately go into convulsions, even if it has been in a warm room for several hours, and has been supplied plentifully with food. Death usually puts an end to its suffering in a day or two."

Such catastrophes are more likely to occur, however, in the spring, after the birds have begun to come northward, than in the steadier weather of January; and even the song-sparrows and snow-birds, which have successfully withstood the rigors of the lowest midwinter temperature, as often succumb as the less inured songsters from the South.

The favorite among our winter birds, perhaps because the most domestic, taking the place of England's robin-redbreast, is the slate-colored snow-bird, which is one of the sparrows. It comes to us with the first frosts, and stays until the wake-robin and spring-beauty bloom. Even then some of them do not go far away to spend the summer, for they breed in the heights behind the Delaware Water-Gap, and also in the Catskills. The main body, nevertheless, go to Canada and Labrador. In the Rocky Mountains I have seen them many times in midsummer as far south as the latitude of Cincinnati; but there the Canada jay also breeds, although in the East its nest is never found—great altitude in the Sierras affording the same climate which eastward is only to be attained at high latitudes.

The nest of the snow-bird is placed on the ground among the moss, or under the protection of the root of a tree, and is built of grass, weed-stalks, and various fibres. The eggs are whitish, sprinkled with pale chocolate and dark reddish-brown. Several species besides our *Junco hyemalis* are found in mountainous parts of the far West and North-

116   FRIENDS WORTH KNOWING.

CARDINAL-GROSBEAK.

west, but they intergrade confusingly, and their nidification is essentially the same. A snow-bird is a snow-bird from one end of the country to the other, and the sharp, metallic note is characteristic of the whole genus.

Truer spirits of the driving snow—for the junco is a sort of fair-weather bird after all—are the snow-buntings, or snow-flakes, or white snow-birds, or, absurdest of all, winter-geese, as the Nahant fishermen call them. Their systematic name is *Plectrophanes nivalis*, and their plumage is handsomely marked with white and chocolate-brown. Sometimes a flock of these buntings will whirl into our door-yard for a brief moment; but in general you must go to the upland fields and frozen marshes to find them, and the best time is just after a "cold snap" or a heavy snow. The Hackensack meadows at such times are full of them, and I have seen flocks of hundreds pirouetting over the ice-covered, wind-swept

A YELLOW-BIRD IN WINTER DRESS.

shores of Lake Erie, or whirling down the bleak sands of Cape Cod. What attracted them to such exposed and dreary spots I could never divine. When they first come they seem unsuspicious of any special danger from man, yet are continually skurrying away from some imaginary cause of alarm. Never going far south of New York, we see few of them even here in mild seasons, and, as the close of the winter approaches, they are among the first to hasten to their home within the arctic circle. In every alternate flock of snow-flakes may perhaps be found one or two Lapland longspurs—another bird which builds its nest in the moss at the foot of Greenland glaciers. Its coat is white and black and chestnut, so that it is easily distinguishable from its lighter fellow, but it is very uncommon.

Next to the diminutive humming-bird, the smallest bird on this continent is the golden-crested kinglet, on whose tiny brow rests a coronet of gold, fiery red and black, below which the jewelled eye is set in a soft, dusky background of olive-green. From tooth to tail he is not so long as your finger, yet this pygmy braves the fury and desolation of winter as cheerily as though soft skies arched overhead. I owe him many thanks for piping his nonchalant, contented little lecture into my ears when I have growled at the weather and the "foolishness" which dragged me out-of-doors on certain terrible days, only to see what such absurd fellows as he were about. He is the most independent, irrepressible little chap I know of, and for the life of me I never can be down-hearted when he is by. In summer the gold-crest (like his royal brother, the ruby-crown) is a fly-catcher, expertly seizing insects on the wing; and on warm days in winter he forages in the tree-tops for such moths and beetles as are abroad; but necessarily he must subsist chiefly on the larvæ which hibernate under the rotten bark, and upon insects' eggs. Thus he is helped to many a meal by the sapsuckers and tomtits, whose stronger bills tear open the recesses where the larvæ lie. In summer the kinglets retreat to boreal regions to rear their young; but we know very little about their domestic life. Just before they leave us in the spring I may, perhaps, have the rare treat to hear a long way off the resonant song of this minute minstrel—bold and clear, carrying me away aloft like that of the English skylark.

Another personification of

"Contented wi' little, and canty wi' mair,"

is the brown creeper, whose bill is curved, and long, and tender, so that he can do very little digging for himself, but follows in the track of the woodpeckers and nuthatches, and picks up the grubs which their vigorous beaks have dislodged, or searches carefully for such small insects, and their eggs, as are not well concealed. There is one now in the tree next my window, in the edge of the city, as I write. He flew from the neighboring horse-chestnut to the foot of the ailantus, and began a spiral march upward. I see him creep steadily round and round and round the trunk, with his tail pressed in against the tree to sustain him (like the pointed stick trailing behind a Pennsylvania wagon), peering into every crevice, poking his bill into all the knot-holes and scars where limbs have been shivered off, running out on each branch, here picking up half a dozen eggs that only a bird's sharp eye could find, there transfixing with his pointed tongue some dormant beetle laid away on his bark shelf, or tearing open the pupa-case of some unlucky young moth, snugly dreaming of a successful *début* in May. This creeper is always to be found in our winter woods and orchards, yet is nowhere abundant; its life is a solitary one, and, although not shy, it is so restlessly active as easily to elude the eye. If, in the early spring, you have the rare fortune to hear its song, regard the privilege as precious.

Another creeping bird, almost always moving head downward, more often seen in midwinter, because then he approaches civilized life, while in summer he retires to the remote woods to rear his brood, is the familiar nuthatch, whose peculiar *nee-nee-nee*—the most indifferent, don't-care-a-bit utterance in the world—is heard from every other tree-trunk. Like the brown creeper, the nuthatches seek their food on the boles of trees, examining every part by a spiral survey—a sort of triangulation—and are not content till the top is reached, when they dive straight to the roots of the next tree, and begin a new exploration. There is no time wasted by these little engineers in foolish flying about or profitless research. Not allowing a cranny to go untouched, they drag out every unhappy grub it shelters before raiding the next hiding-place of insect-life. Their feet are broad and strong for clinging; their bills are small pickaxes, their tongues harpoons, and their brains marine clocks, just as steady one side up as another. Thus they are able to live on the injurious borers and the like which pass through their metamorphoses beneath the bark; and, except when everything is incased in ice, do not eat seed, or even alight on the ground. They are among the most active and serviceable of the fruit-grower's benefactors, continuing, during the cold months, the good work dropped in October by the summer birds, and finding in his insidious enemies their favorite food. The nuthatch is the leader of that admirable little company composed of the chickadee, the crested titmouse, the downy woodpecker, and sometimes of the red-bellied nuthatch and *spirituel* creeper, which Wilson truthfully describes as "proceeding regularly from tree to tree through the woods like a corps of pioneers; while, in a calm day, the rattling of their bills, and the rapid motions of their bodies, thrown like so many tumblers and rope-dancers into numberless positions, together with the peculiar chatter of each, are altogether very amusing, conveying the idea of hungry diligence, bustle, and activity."

Every one knows the black-capped titmouse—our jolly little chickadee, and his jolly little chant:

> "*Chick-chickadeedee!* Saucy note,
> Out of sound heart and merry throat,
> As if it said: 'Good-day, good sir!
> Fine afternoon, old passenger!
> Happy to meet you in these places,
> Where January brings few faces.'"

He is the hero of the woods; there are courage and good-nature enough in that compact little body, which you may hide in your fist, to supply a whole groveful of May songsters. He has the Spartan virtue of an eagle, the cheerfulness of the thrushes, the nimbleness of the sparrow, the endurance of the sea-birds, condensed into his tiny frame, and there have been added a "peartness" and ingenuity all his own. His curiosity is immense, and his audacity equal to it; I have even had one alight upon the barrel of the gun over my shoulder as I sat quietly under his tree. The chickadees come to us with the first frost; and keen eyes may discover them all the year round in the Catskills, or among the heights of the upper Delaware River, whither they go to nest, the majority, nevertheless, passing to Canada for that purpose.

There is a winter wren also, but, although considerably smaller, it is frequently mistaken for the inquisitive and saucy house-wren, which fled south in October. It is a species heard rather than seen, evading observation in the dense brush, through which it moves more like a mouse than a bird. Its prolonged and startling bugle-song is a wonder, and its whole history is charming, but I must pass it by. If you wish to become acquainted with him (and several of his midwinter associates) in more genial days, you have only to go to the depths of the Catskills or Adirondacks, where he spends his summer.

The family of sparrows, finches, and buntings—the *Fringillidæ*—supplies more of the winter woodland birds than any other single group, the list of those regularly present in January including the pine-grossbeak, the red and the white-winged crossbills, the two red-poll linnets, the pine, grass, and gold finches, the song, tree, and English sparrows, besides an occasional straggler like the purple finch, cardinal, and white-throat. The first five mentioned are polar bred, and return to their native heaths at the earliest intimation of spring. The pine-grossbeak is a big, clumsy-looking bird, with a plumage reminding you of a blossoming clover-field—a mixture of red and dull green. It has found out what its thick, strong bill was made for, and crushes the scales of the tough pine-cones as though they were paper. The pine-grossbeaks often come into the vil-

CROSSBILL.

lage streets, hopping about in search of almost anything to eat, and are very tame and interesting. Their note is a cheery. one, and when captured they thrive well in the cage, eat apple-seeds greedily, and become very entertaining. The pine-finch, or siskin, is its miniature, and seeks much the same sort of food, but must get it from softer cones, for its bill does not seem half as stout. It is erratic in its visits, and its actions outside of the pine-trees are precisely like those of its cousin, the yellow-bird.

All winter you may notice along the field-fences and in the grassy plats beside the railway, where weeds have gone to seed, active flocks of small, plainly-attired little birds, as cheerful as can be. These are our thistle-loving goldfinches, or yellow-birds, whose simple, sweet song and billowy flight were part of the delight of last summer, but which now have exchanged their gay livery of canary-yellow and black for sober undress suits of Quaker drab. The goldfinches, as such, appear with the apple-blossoms, and are seen no later than the gathering of the fruit; but their seeming disappearance in autumn, and reappearance in spring, are only changes of plumage. Nevertheless, they are not so abundant in winter as in summer, many moving a little distance southward. The crossbills are naturally so named, for the tips of their mandibles slide by one another instead of shutting squarely together. Whether or not this peculiarity has been gradually acquired to meet the necessity of a peculiar instrument to twist open the cones and other tough pericarps, upon the contents of which they feed; or whether it is an accident perpetuated and made the best of; or whether the crossed bill was "created" in that fashion in the beginning, with a definite intention toward pine-cones, we may theorize upon to suit our tastes: but certain it is that it answers the bird's purpose most admirably. The red crossbill is the more common of the two, but the white-winged is not greatly different. They fly in small flocks, often coming among the gardens, where their odd appearance never fails to attract attention. In addition to pine-seeds, they feed on the seeds and buds of the cedar, birch, alder, mountain-ash, Virginia creeper, etc., and probably add apples, haws, and berries to their bill of fare, as does the grossbeak. They are wonderfully happy creatures, fluttering in and out of the evergreens, or passing swiftly from one to another, working away at a swinging cone "teeth and toe-nail," heads or tails up—it doesn't matter—till every kernel is extracted, then with one quick impulse launching into the air and departing— perhaps for the arctic circle—before you have had time to bid them good-bye.

One of the earliest and handsomest migrants from the frozen North is the little red-poll linnet, which is about the size of a stout canary. He is a dandy, changing his gay suit of black, brown, white, saffron, pink, red, and crimson several times a year, and—at least until he is three or four years old—never dressing twice alike. He is an exceedingly melodious if not a very versatile singer, in England is often kept in cages and mated with the canary, and might be here. There would be no difficulty in catching him.

Two other of the familiar friends who make our spring meadows vocal with an incessant concert, the song-sparrow and grass-finch, remain with us through the winter also; but more than half the song-sparrows are frightened southward by the first snow-storm. A few, however, are always to be met with in the swamps and edges of the woods during January, living under cover of the briers and brush-heaps, and upon the seeds of various grasses and herbs, scratching up the leaves to get at dormant insects or their eggs, here picking up a checker-berry which the snow has not drifted over, there nibbling at the dried remains of blackberries, raspberries, and wrinkled crab-apples, squeezing the gum from a swelling bud, tearing open the seed-case of the wild-rose whose blossom they shook to pieces as they darted to their nests in early June. The brown grass-finch—easily recognized by the two white feathers shown in the tail when flying—seems scarcely ever to leave the field in which it was born. It is emphatically a bird of the meadows, where its song is heard loudest in the long summer twilights when most other birds are silent, so that Wilson Flagg called it the vesper sparrow. Building its nest in a little hollow on the ground, finding its food among the grass, it seems hardly to fly over the boundary-fence from one year's end to another. How these finches are able to stand the winter in the open fields, is a mystery; perhaps they go elsewhere at night, or crawl into holes; but you may meet them scudding across the uplands every month of the year, keeping company with the few meadow-larks that remain.

All this month, in hedge-rows, wooded hollows, and thickets, beside springs of water, where very likely you may flush a woodcock, will be heard the low warble of the tree-sparrows, northern cousins of the trilling chippy of our lilac-bushes, and of the pretty field-sparrow that from every green pasture calls out, *C-r-e-e-p, c-r-e-e-p, c-r-e-e-p, catch'm, catch'm, catch'm!* as my mother used to phrase it for me. They receive the name from the habit of taking to the trees when disturbed, instead of diving into the bushes and skulking away as do the other sparrows; but the less common name, Canada sparrow, is better. Once in awhile they come into the towns: I saw one yesterday in the horse-chestnut in front of my window, which seemed to be finding plenty to eat about the bark and scanty leaves that remained, until the English sparrows got news of his presence and drove him away in their buccaneering style.

These same outrageous English sparrows are the most conspicuous, really, of all our January birds. They are spreading widely through the suburbs of the city, especially between here and Philadelphia; and I am sorry to see it, for they are uncompromising enemies to all our native birds.

It would lead me to far overstep the reasonable limits of this essay if I attempted to extend to all the winter birds even the brief sketch I have given of some of the woodland species. A mere mention must suffice.

Some birds besides those already noticed are residents with us the year round: thus a few robins, bluebirds, crows, bluejays, cedar-birds, kingfishers, flickers, blackbirds, purple finches, wild pigeons, quails, grouse, and woodcocks, are always likely to be found in the neighborhood of New York in January; while one or two of the arctic woodpeckers, the Canada jay, the waxwing, and some other rarities, may be met with at long intervals. Of the birds of prey, we have in greater or less numbers this month the golden and bald eagles (about the Palisades), an occasional osprey, the rough-legged, red-shouldered, and red-tailed buzzards, the marsh-harrier, and some others; and, among owls, the fierce snowy owl, which will take a grouse from its roost, or carry off a hare; the barred, great horned, long-eared, short-eared, mottled, and little saw-whet owls. Along the adjacent shores of Long Island and New Jersey are seen the various sea-ducks, "coots," and geese; the loon, and an occasional northern sandpiper, like the splendid purple one; the herring, kittiwake, laughing, black-backed, and several other gulls; and irregularly certain wandering

THE WAXWING.

sea-birds whose lives are not so much affected by climatic conditions as are those of the land-birds.

Deprived of the small reptiles, the young of squirrels and other mammals, eggs, and the large night-flying moths and beetles which in summer form a good portion of their subsistence, the predaceous birds become more fierce in winter than at any other time, and exercise all their cunning in the pursuit of such meadow-mice and other animals as are imprudent enough to step out of their subnivean galleries, or in the capture of weaker birds. The few late fish-hawks remain by the sea-shore, plunging in now and then for their finny prey, which the bald eagle very often compels them to relinquish to him. The golden eagle, covering the landscape with keen and comprehensive glance as he sweeps over in vast circuits, swoops upon hares, foxes, and the like, sometimes even picking up an early lamb, or catching a grouse before it can baffle its dreaded pursuer by burying itself in the snow. The buzzard and marsh-hawk sail low over the meadows in slow and easy flight, or stand motionless above some elevated spot in the lowlands, watching intently until a mole, or shrew, or mouse, shows itself below, when they drop upon it like a shot, and carry it off before the poor victim has time to recover from its palsy of terror. Less frequently do these species seem to catch birds, and between Christmas and Easter they lead a very precarious existence. The owls, too, must "live by their wits," but, being nocturnal, they have the advantage of the birds, and, we may be sure, snatch many a tender one rudely from its roost in the open trees, although the dense twigs and sharp needles of the cedars and other close-boughed evergreens must offer such obstacles to the rapid

132 *FRIENDS WORTH KNOWING.*

passage of the owl as to allow many an intended victim to escape. The larger species, as the farmer well knows, will

UNWELCOME!

often in continued cold weather come into the very barnyard and carry off his chickens; while the nocturnal habits of most of the smaller mammals not hibernating in January lead them abroad when the owls are mostly flying, and on moonlight nights these prowlers get many a good meal, no doubt.

It would seem, therefore, as if the chances of death presented to the lesser winter birds by scarcity of food, rigor of climate, hawks by day and owls by night, outnumbered the chances of life offered by their alertness and enduring vitality. But there are some additional circumstances favorable to their escape from the latter fate, their resources against starvation and freezing having already been explained. One of these circumstances is the vigilance of the birds: they never are forgetful. Sometimes their curiosity leads them into danger, or an enemy like man, which they do not suspect, may approach them by being very quiet; but a hawk could never insinuate himself into a sparrow's good graces, nor could an owl win his confidence; both must trust to surprising him or overtaking him in an open race, which is about as difficult as "catching a weasel asleep." Then the hiding-places of the birds in hollow trees, crannies in walls, dense thickets, and brush-piles, during the night and in bad weather, are such as afford excellent security from their nocturnal winged enemies, although quite accessible to foxes and weasels. It is a curious fact that fourteen or fifteen of our January birds choose hollows in trees or holes in the ground for nesting-places, as though consciously profiting by their experience of the security afforded.

Another very important circumstance favoring the preservation of small birds at this season is the fact that in the majority of cases the tints of their plumages are precisely such as best harmonize with the surroundings in which they are most often seen, and thus make them less discernible than they otherwise might be. Looking through our list of winter birds, many striking examples of this protective coloration are found—more, in proportion, than in summer, when there does not seem to be so great need of individual safety, and the "struggle for existence" is not narrowed down to such a strait, and beset with so many difficulties. The kinglets, for instance, spend their time in flitting about the tops of the trees, and their plumage is found to be a dusky green, like an old leaf, while the fiery crowns which both wear are concealed, except at moments (of love-passion, I imagine) when they wish to display them. Easier to detect than the kinglets, yet plainly dressed, are the titmice and nuthatches; but these frequent widely different scenes, and, moreover, have compensating advantages beyond most other birds in the habit of living mostly in the deep woods where diurnal birds of prey are uncommon, and at night of secreting themselves in small holes where the owls cannot get at them. This is also true of the small spotted woodpeckers, which, nevertheless, are very inconspicuous objects upon the dead and white trunks they frequent.

The brown and white streaks of the creeper (*Certhia americana*), however, seem to me to furnish a decided case of protective colors in plumage, since they harmonize so exactly with the rough, cracked bark along which the creeper glides, that the wee bird is hardly to be followed by the eye at a moderate distance. Again, no coat would better help the wren to scout unobserved about the tangled thickets and through the piles of wind-drifted leaves in and out of this and that shadowy crevice than the plain brown one he wears; while the lighter tints of the goldfinch's livery are precisely those which agree with the russet weeds and grass whose harvest he diligently gathers. The group of exclusively boreal birds seems especially protected from harm by the correspondence of their coat and their surroundings. Their home is among the evergreens, where an occasional dead branch or withered stem relieves the verdancy with yellowish patches, and the thick-hanging cones dot the tree with spots of reddish-brown; their plumage is mottled with green, tints of yellow and brown, an inconspicuous red, and a little black and white—just the colors one's eye takes in at a glance as he looks at a hemlock. The practical result for our eyes (or a falcon's) is, that the pine-grossbeaks and finches, the crossbills and purple finches, blend with the foliage and cones and dead branches until they are lost to any but the most attentive gaze. The snow-bunting rejoices in a cloak of white, and thus mingles inextricably to the eye with the feathery flakes he whirls among, while his companion, the longspur, is almost equally ghostly. All the winter sparrows are of the brown color of the sere grass, withered leaves, and broken branches among which they dwell, except the slaty snow-bird, and he is of a neutral tint, easily lost to view in a shadow.

This protection of adaptive colors is not enjoyed to any great extent by the robin, bluebird, meadow-lark, cardinal-grossbeak, and kingfisher—but none of these are "winter" birds here, properly speaking, but only loiterers behind the summer host, and ought really to be excluded from the comparison; nor by the crow, crow-blackbirds, bluejay, Canada jay, and butcher-bird—but these are all large and strong, able for the most part to defend themselves; while, on the contrary, the colors of the large but timid and defenceless woodcock, quail, and grouse are highly protective.

Birds of prey themselves scarcely need such protection from one another, yet some of them regularly exchange their summer plumage for a winter dress of lighter and (in the general white of the landscape) less conspicuous tints; but this may operate to their advantage in the reverse way of allowing them to attain a closer, because unobserved, approach to their quarry. This leaves us, among the land-birds, only the bright red-poll and the waxwings as exceptions to the supposed rule that the plumages of winter birds are colored in a way directly favorable to their special preservation at that season of augmented danger. They are cases of which I have no account to give other than that—let me beg the reader charitably to believe—these are the exceptions which "favor the rule."

But against one persecutor no concealment of natural color or artful device avails, and the brains of the pretty songsters, so full of wit to avoid other enemies and provide for each day's need, are his choice repast. This dainty tyrant wears an overcoat of bluish ash trimmed with black and white, a vest of white marked with fine, wavy, transverse lines, white knee-breeches, and black stockings. His eyes are dark and piercing; his nose Napoleonic; his forehead high and white; his mustache as heavy and black as that of any cavalier in Spain. This Mephistopheles among birds is a ruffian, truly, yet with polish and a courage without bravado which commend him. Being an outlaw in the avian kingdom, he can only maintain himself by adroitness and force, but has such singular impetuosity, prudence, and fortitude, that he is not only able to keep himself and his retainers in health and wealth and happiness, but to gratify his blood-thirsty love of revenge by killing numberless innocents without mercy. Thus he has struck terror to the heart of every feathered inhabitant of the January woods. Like Cæsar, he knows and joyously endures hunger and cold and thirst. Is it biting, freezing weather, and blinding snow? Little cares he; he can then the more easily surprise his benumbed prey. Is it a warm, sap-starting, inviting day? He is at the festival of the birds — a fatal intruder into many a happy circle. His favorite perch is the high rider of some lonely fence, where he quietly waits till a luckless field-mouse creeps out and he is able to pounce upon it; or an incautious sparrow or kinglet dashes past, unconscious of the watchful foe who seizes him like a flash of lightning. Having felled his quarry with a single blow, he returns to his fence-post and eats the brains —rarely more—or perhaps does not taste a single billful, but impales the body upon a thorn, or hangs it in an angle of the fence, as a butcher suspends his quarters of beef. It used to be thought this murderer thus impaled nine captives, and no more, so he was christened nine-killer; the book-men labelled him *Collurio borealis;* we know him as the butcher-bird: he is the arctic brother of the shrikes, and the boldest, bravest, noblest, and wickedest of his savage race.

A SHRIKE.

VI.

*THE BUFFALO AND HIS FATE.**

PERHAPS no indigenous animal of this country has attracted more attention or met with a greater number of biographers than the bison or buffalo. Its history has been a tale of extermination, and a very short time will be likely to see the last of these noble beasts roaming over the plains. For hundreds of years a small remnant of the ancient herds of aurochs, the native European bison, have been preserved in the parks of the nobility; but in this "free" country not even this means of safety seems left to our persecuted buffalo.

To the Spanish colonists the American bison was commonly known under the name of *civola*, while the French usually called it *le bœuf, buffle, vache sauvage*, or *bison*

---

* A review of Prof. J. A. Allen's "The Bison, Past and Present, in this Country," forming Part II. of Volume I. of the "Memoirs of the Geological Survey of Kentucky," Prof. N. S. Shaler, Geologist, in charge; also reprinted by the Museum of Comparative Zoology as one of its "Memoirs." Cambridge, 1875.

A MOTHER AND CHILD OF THE PLAINS.

d'Amérique. Peter Kalm, who travelled through America in 1749, spoke of them as *wilde ochsen* and *kühe.* But the word *buffalo*—at first spelled *buffelo*—soon replaced the earlier names. Scientific men claim that our species (*Bison americanus,* Smith) should be called bison, as "buffalo" is applicable only to the East Indian genus *Bubalus.*

It appears that our bison has already outlived at least two other races, which exceeded it in size—the *Bison latifrons* and the *Bison antiquus.* The former was contemporary with the mastodon, and was an ox of gigantic bulk, the tips of whose horns were eleven or twelve feet apart, and which probably stood as high as an elephant. Of the latter species more abundant remains have been dug up, particularly from the ice-cliffs at Escholtz Bay, on the Arctic coast north of Alaska. This fossil ox was of smaller size than the *Bison latifrons,* but much larger than the existing buffalo, although not greatly different from it in form. It seems to have been spread over the northwestern half of the continent from the Ohio Valley to Alaska, its remains occurring everywhere with those of the larger extinct mammalia, yet it may have survived to a comparatively recent date.

With the appearance of the buffalo, which only a few decades ago swarmed in prodigious herds over nearly a third of North America, all are familiar. The male measures about nine feet from the muzzle to the insertion of the tail; the female about six and a half feet. The height to the top of the hump of the male is five and a half to six feet, and of the female about five feet, sloping in each case to a height at the hips of four and a half to four feet. The weight of the old males is nearly two thousand pounds, while the cows weigh one thousand to twelve hundred pounds. The horns are short, thick at the base, curved, and sharply pointed; the hoofs are short and broad; the short tail ends in a tuft of long hairs. In winter the head and whole under parts are blackish brown; the upper surface lighter, fading as spring advances. Young animals are of a darker, richer brown than the old ones, age bleaching the thick masses of long, woolly hair, which falls so abundantly over the shoulders and face, to a light yellowish-brown. In the spring the hinder parts are almost naked through the moulting of the hair, while that upon the shaggy fore parts remains permanently. Pied coats are occasionally met, and examination and measurements of skulls and skeletons show much individual variation in form and proportions.

As is well known, the buffalo is pre-eminently gregarious —herds numbering millions of individuals, and blackening

TRAVELING HERDS.

the whole landscape, having formerly been met with constantly on the plains. Emigrant trains used to be delayed by the passing of dense herds, and during the first years of the Kansas Pacific Railway its trains were frequently stopped by the same cause. These masses seem to have some sort of organization, consisting of small bands which unite in migration or when pursued, but separate when feeding. The cows, with their calves and the younger animals, are generally toward the middle of the small herd, while the older bulls are found on the outside, and the patriarchs of the herd bring up the rear. Much romancing has been wasted on this simple and natural grouping by writers who have described the supposed regularity and almost military precision of their movements. The sluggish, partly-disabled old males constitute the "lordly sentinels" of such tales, who are supposed to watch with fatherly care over the welfare of their "harems." The truth is, that these protectors, fancied so alert, are the most easily approached of any of the flock, and the real guardians are the vigilant cows themselves, who usually lead the movements of the herd.

The rutting-season is July and August. The period of pregnancy is nine months, and rarely more than a single calf is born, which follows the mother for a year or more.

During the rutting-season the bulls wage fierce battles, but they rarely result fatally. The short horns are not very dangerous weapons, and the masses of hair on the forehead break the force of the stunning collisions. At this season the bulls become lean, regaining their flesh in autumn, while the cows are fattest in June. During its moulting in midsummer the animal possesses a very ragged and uncouth appearance, the hair hanging here and there in matted, loosened patches, with intervening naked spaces; and it endeavors to free itself from this loosened hair by rubbing against rocks and trees, or rolling on the ground. The coats are in prime condition for robes in December.

The buffalo is nomadic in its habits, roaming in the course of the year over vast areas in search of food or safety. The fires that annually sweep across thousands of square miles of the grassy plains, the ravages of grasshoppers, often destroying equally extensive tracts of vegetation, and the habit of keeping in compact herds, which soon exhaust the herbage of a single region, all compel constant movement. There is a popular belief that the buffaloes used to migrate from the northern plains to Texas in fall and back again in spring, but this seems erroneous. Before the intersection of the West by railroads and emigrant trails their movements were more regular, no doubt,

THE SIGNAL.—BUFFALO HERD IN SIGHT.

than at present, and slight northward and southward migrations are well attested as occurring in Texas and also on the Saskatchewan plains; but the herds constantly winter as far north as the latter region, and for twenty-five years have not passed southward even to the Platte. In the extreme north they leave the exposed plains in winter and take shelter among the wooded hills. Such local movements as these were formerly very regular, and hunters knew just where to look for their game at any season of the year.

The behavior of the buffaloes is very much like that of domestic cattle, but their speed and endurance seem to be far greater. When well under way it takes a fleet horse to overtake them, and they raise a column of dust which marks their progress when miles away. They swim rivers with ease, even amid floating ice, and show a surprising agility and expertness in making their way down precipitous cliffs and banks of streams, plunging headlong where a man would pick his way with hesitation. Ordinarily, however, the buffalo exhibits commendable sagacity in his choice of routes, usually taking the easiest grades and the most direct course, so that a buffalo-trail—often worn deep into the ground—can be depended on as affording the most feasible road through the region it traverses.

When belligerent, the old bulls make the most blustering demonstrations, but are really cowardly. Facing the approaching hunter with a boastful and defiant air, they will pace to and fro, threateningly pawing the earth, only to take to their heels the next moment. The bulls greatly enjoy pawing the earth and throwing it up with their horns, digging into banks or getting down upon one knee to strike into the level surface, so that the sheaths of their horns are always badly splintered. They are very fond, too, of rubbing themselves, and evidently regard the telegraph-poles along the railroads as set there for their especial convenience in this respect. A line of telegraph was built between Helena, Montana, and Fort Benton. But it was found impracticable to maintain it beyond Fort Shaw, where the mountains end, and when I passed there in 1877 the attempt had been abandoned. The buffaloes pushed the poles down, and then getting entangled in the wire, broke it to pieces. Fragments of this wire, twisted about their horns, were carried many miles, and are still occasionally picked up by hunters all over the grassy uplands that stretch so boundlessly northward from the upper Missouri.

But their chief delight is "wallowing." Finding in the low parts of the prairie a little stagnant water among the grass, or at least the surface soft and moist, an old bull plunges his horns into the ground, tearing up the earth and soon making an excavation into which the water trickles, forming for a short time a cool and comfortable bath, in which he wallows like a hog in the mire, swinging himself round and round on his side, and thus enlarging the pool until he is nearly immersed. At length he rises besmeared with a coating of mud, which, drying, insures him immunity from insect pests for many hours. Others follow, each enlarging the "wallow-" until it becomes twenty feet in diameter, remains a prominent feature in the landscape, and forms a cistern where a grateful supply of water is often long retained for the thirsty denizens of that dry region, both animal and human.

Like the other species of the bovine group, the bison is of sluggish disposition, and mild and timid, ferocious as his shaggy head and vicious eye make him look. He rarely attacks, except in the last hopeless effort of self-defence. "Endowed with the smallest possible amount of instinct," says Colonel R. I. Dodge, "the little he has seems adapted rather for getting him into difficulties than out of them. If not alarmed at sight or smell of a foe, he will stand stupidly gazing at his companions in their death-throes until the whole herd is shot down. He will walk unconsciously into a quicksand or quagmire already choked with struggling, dying victims." Having made up his mind to go a certain way, it is almost impossible to swerve him from his purpose, and he will rush heedless into sure destruction. Two trains were "ditched" in one week on the Atchison, Topeka, and Santa Fe Railroad by herds of buffaloes rushing blindly against and in front of them. Finally the conductors "got the idea," and gave the original occupants of the soil the right of way whenever they asked it. During a voyage that I made down the upper Missouri in 1877, our steamer more than once had to stop to allow swimming herds to get out of the way, and once we completely keelhauled a sorry old bull. Yet, as Mr. Allen suggests, their inertness may be exaggerated by writers, as their sagacity certainly has been. This stupidity, unwariness, or liability to demoralizing panic, places them at the mercy of the hunter, who is their only enemy besides the wolves. In former times, young or weak animals straying from the herds, and all the wounded and aged that could be separated from their fellows, were quickly set upon and worried to death by wolves; but now these brutes have become so reduced as not to form a serious check upon their increase.

The early explorers of the Mississippi Valley believed that the buffalo might be made to take the place of the domestic ox in agricultural pursuits, and at the same time

INDIANS KILLING BUFFALOES ON THE UPPER MISSOURI.

yield a fleece of wool equal in quality to that of the sheep; but no persistent attempts have yet been made to utilize it by domestication. That the buffalo-calf may be easily reared and thoroughly tamed has been conclusively proved, but little attention has been paid to their reproduction in confinement, or to training them to labor. During the last century they were domesticated in various parts of the colonies, and interbred with domestic cows, producing a half-breed race which is fertile, and which readily amalgamates with the domestic cattle. The half-breeds are large, fine animals, possessing most of the characteristics of their wild parentage. They can be broken to the yoke, but are not so sober and manageable in their work as the tame breed—sometimes, for instance, making a dash for the nearest water, with disastrous results to the load they are drawing. It is somewhat difficult, also, to build a fence which shall resist the destructive strength of their head and horns. But the efforts at taming buffaloes have not been many, or seriously carried on, and no attempt appears to have been made to perpetuate an unmixed domestic race. Probably after a few generations they would lose their natural untractableness, and when castrated would doubtless form superior working-cattle, from their greater size, strength, and natural agility.

"The fate of extermination so surely awaits, sooner or later, the buffalo in its wild state, that its domestication becomes a matter of great interest, and is well worthy the attention of intelligent stock-growers, some of whom should be willing to take a little trouble to perpetuate the pure race in a domestic state. The attempt can be hardly regarded otherwise than as an enterprise that would eventually yield a satisfactory and profitable result, with the possibility of adding another valuable domestic animal to those we now possess."

The precise limit of the range of the buffalo when the first Europeans visited America is still a matter of uncertainty, yet its boundaries at that time can be established with tolerable exactness. It was beyond doubt almost exclusively an animal of the prairies and the woodless plains, ranging only to a limited extent into the forested districts east of the Mississippi River. The results of the present exhaustive inquiries seem to show that its extension to the northward, east of the Mississippi, was limited by the Great Lakes. Contrary to the supposition of several recent writers, Mr. Allen has not been able to find a single mention of its occurrence within the present limits of Canada, New England, or New York State, although the name of the city of Buffalo and the neighboring "Buffalo Creek" prob-

A FIGHT AGAINST FATE

ably imply that this animal once extended its travels to that point. All the supposed references to its being seen on the St. Lawrence, or in Canada West, turn out to mean the elk—the same indefinite terms being often used for both by early writers—or else to apply to some part of the broad territory then called Canada, but not now included within its limits. Changes in political boundaries have constantly to be borne in mind in studying ancient narratives.

Furthermore, no remains of the bison have been found among the bones in the shell-heaps along the Atlantic coast, and there is no unquestionable evidence, among all the early lists of the natural products of the country, of its occurrence anywhere on the seaboard north of the Potomac for a long period preceding the discovery of the continent by Europeans. The only well-authenticated instances of its being found east of the Blue Ridge are the apparently casual passage of small herds through the mountains from West Virginia into the upper parts of North and South Carolina by way of the New, Holston, and French Broad rivers. They seem to have been common on the savannahs about the heads of rivers in the western parts of those states; but it is well attested that they never came down to the sea-coast. Nor can good evidence be shown that they ever reached any part of Georgia, Florida, or Alabama (although possibly Mississippi), as at present bounded, not appearing habitually to have penetrated south of the Tennessee River—unless just along the bank of the Father of Waters—on account of the thickness of the forest.

The records in general, then, show that at the beginning of the seventeenth century the range of the buffalo east of the Mississippi, with the exception of its occasional appearance on the eastern slope of the Alleghanies in the Carolinas and Virginia, was restricted to the area drained by the Ohio River (except over the lowlands at its mouth), and to the eastern tributaries of the Mississippi in northern Wisconsin and Minnesota; also that it was very numerous, and uniformly distributed over the prairies of Illinois and Indiana, and also about the upper tributaries of the Ohio; but less numerously and uniformly over Ohio, West Virginia, Kentucky, western Pennsylvania, and the northern portion of Tennessee, being everywhere restricted to the prairies and scantily wooded land along the streams.

In the appendix to Mr. Allen's admirable monograph, Professor N. S. Shaler offers a short discussion of the probable age of the bison in the Ohio Valley. In the swamps surrounding the "salt-licks" of Kentucky, buffalo-bones are found packed in great quantities in the mucky soil, but only about the latest vents of the saline waters, which have from time to time changed their points of escape from the ground. The caverns of Kentucky and Tennessee, which were the homes of the aboriginal people of the region, and receptacles for their dead, and where have been found skeletons of the beaver, deer, wolf, bear, and many other mammals, have never yielded any bones of the bison. Moreover, among all the many figures of animals and birds found on the pottery and ornaments of the prehistoric races of the West, the marked form of the buffalo does not appear, making it presumable that this animal was unknown to the people who built the mounds. Professor Shaler is of the opinion, held by many ethnologists, that the "mound-builders" were essentially related to the Natchez group of Indians, and were driven southward by ruder tribes of red-men from the north and north-west. The Indians north of the Ohio are known to have been much in the habit of burning the forests, and no doubt the invaders alluded to above signalized their advance by such conflagrations. This making of plains by the repeated burning of forests, aided by "the continued decrease of the rainfall, which was a concomitant of the disappearance of the glacial period," permitted the buffalo to advance rapidly eastward as far as the Alleghanies, and, coincidently, as far as the mound-building people appear to have settled the country. Its advent thus seems to have been singularly recent.

The question of the origin of the buffalo and its relation to the earliest tribes of people in the Ohio Valley is made still more complicated by the fact that an earlier and closely related species of buffalo, probably coeval with the mammoth and musk-ox, and possibly with the caribou and elk, was living at the time just following the close of the glacial epoch. "I am strongly disposed to think," writes Professor Shaler, "that in the *Bison americanus* we have the descendant of the *Bison latifrons*, modified by existence in the new conditions of soil and climate to which it was driven by the great changes closing the last ice age." But he adds that future explorations will probably show that there was an interval of some thousands of years between the two species along the Ohio.

Although the main chain of the Rocky Mountains has been supposed commonly to form the western limit of the range of the buffalo, there is abundant proof of its former existence over a vast area westward, including a large part of the Utah basin, the Green River plateau, and the plains of the Columbia, as far as the Blue Mountains of Oregon and the Sierra Nevada. Evidence of this is found in the bleached skulls, in accounts of early explorers, and in traditions of the Indians. During the very severe and snowy winter of 1836–'37 large herds were lost through starvation; by 1840 it had retreated eastward to the forks of the Yellowstone, and been extirpated in the Utah Valley and about the head-waters of the Colorado; and ten years later was never to be found west of the Rocky Mountains, between the British possessions and the Rio Grande del Norte. Westward of this great river it does not seem, within the past two centuries, to have extended itself at all into the highlands of New Mexico; but, farther south, there is proof of its former range over the northeastern provinces of Mexico to at least the twenty-fifth parallel, though it was never abundant there, and abandoned that region before the beginning of the current century.

The great centre of buffalo-life in ages past was the vast expanse of treeless plains which stretch uninterruptedly from the Texas coasts almost to the Arctic Circle, and here, in restricted areas, they have been able to survive until the present time.

When Cabeça de Vaca met the buffaloes in 1530 they ranged throughout nearly all Texas, the higher prairie-lands of north-western Louisiana and Arkansas, and thence uniformly northward and westward. But soon after 1820 they disappeared altogether from Arkansas, and were not seen in western Missouri and southern Iowa later than 1825; but immense herds still roamed over the northern half of the latter State. Since 1845, however, few have been seen anywhere within Iowa, nor did they linger many years longer in Minnesota.

The stream of emigration across the plains to California about 1859 had a curious and permanent effect on the buffaloes. The overland route followed up the Kansas and Platte rivers, and thence westward by the North Platte to the South Pass. The buffaloes were soon all driven from this line of travel; and the great herd which had stretched from the Rio Grande to the Saskatchewan was permanently divided into two — a northern and a southern herd — which were more and more widely separated by the construction of the Union Pacific Railroad. Year by year since, the limits of the range of each division have been contracting under relentless persecution and the encroachments of civilization, until now they are easily circumscribed. The poor beasts have been hunted by the Indians, have been followed incessantly by white men—professional hunters, sportsmen, hide-seekers, and soldiers, who have been afforded easy access to their haunts by the railroads that have penetrated to their ancient pastures, and been given the means of keeping up the hunt by the

THE BITTER END.

nearness of the frontier settlements to the resorts of each herd. Enormous destruction has ensued in Kansas and Colorado, and has had the effect to drive the southern division southward and south-westward into Texas, where hunters cannot or (on account of Indians) dare not follow them. They are, therefore, just now afforded a temporary rest from persecution; but, unless legal interference be quickly made and strict regulations rigorously enforced, the fate of the buffalo south of the Platte will be a repetition of its history east of the Mississippi—speedy extermination.

As to the northern herd, while twenty years ago buffaloes were accustomed to frequent the whole region between the Missouri River and the forty-ninth parallel, from the western boundary of Dakota to the Rocky Mountains and even far into their valleys, they are now restricted to the comparatively small area drained by the southern tributaries of the Yellowstone, and northward over the most of Montana to the Missouri. North of the Missouri River almost a separate subdivision of the herd seems to exist, which feeds between longitude 106° and the Rocky Mountains, and northward to the wooded region of the Athabasca and Peace rivers. Within thirty years they have become extirpated over half of this fertile region north of our boundary, and their numbers, probably, have correspondingly decreased.

It thus appears that in three-quarters of a century the buffalo has been compelled to relinquish a habitat covering a third of the continent for two regions not greater together than the present Territories of Montana and Dakota; and they were formerly just as numerous over the whole extent as they now are in favored spots within their range. Hence the theory that they have not been so much reduced in numbers as they have been circumscribed in range and concentrated upon narrow limits, will not hold good. Over much of this great region they were actually killed on the spot, not driven out.

VII.

THE SONG-SPARROW.

THE American song-sparrow is a peculiar lover of old fields where Nature is fast reasserting herself after the temporary rule of man. The tumbling, lichen-patched stone fences; the gray cattle-paths diverging from the muddy bar-way to those parts of the pasture where the grass is sweetest; the weedy banks of the sluggish brook winding indolently among mossy bowlders and tangled thickets and patches of fragrant herbage—are all congenial to it, and are its chosen resort. Yet it is so common throughout most of the United States that you may find it almost anywhere—skulking about the currant and raspberry bushes in the village gardens; taking a riotous bath in some pool by the roadside, about whose rim, perhaps, the ice still lingers; hastening to the top of a forest-tree to plume its dripping feathers, and shake off at once the crystal water and a crystal song.

Our favorite is the very first bird to greet us in the

spring—in fact, many remain through the winter as far north even as Boston and Lake Erie. It is thought by ornithologists, however, that the winter song-sparrows are not the same individuals that were with us in summer, and which have gone southward, but are inhabitants of more northern latitudes, that have come down with the snow-birds; and it is said that these are far hardier birds, better and more versatile musicians.

During the winter the song-sparrow remains, quiet and busy, along the edges of the woods on warm hill-sides in company with the spotted woodpeckers and snow-birds, or associates with the fowls in the barn-yard for a share of the housewife's bounty. But as the March snow melts, and the sun sends genial warmth to awaken the buds, he mounts the topmost twigs of the brush pile whose labyrinths he has spent the winter in exploring, and pours forth a rapturous welcome to the couriers of summer. Then through all the spring days, whether they be shady or sunny, from early morn till long after sunset, are heard the sweet and cheery cadences of his song, trilled out over and over again like a canary's. He starts off with a few low rattling notes, makes a quick leap to a high strain, ascends through many a melodious variation to the key-note, and suddenly stops, leaving his song to sing itself through in your brain. To amplify another's illustration, it is as though he said "press-*press*-press-by-TEEEE-rian-*ian!*" His clear tenor, the gurgling, bubbling alto of the blackbirds, the slender purity of the bluebird's soprano, and the solid basso profundo of the frogs, with the accompaniment of the April wind piping on the bare reeds of winter, or the drumming of raindrops, form the naturalist's spring quartette—as pleasing, if not as grand, as the full chorus of early June.

The song of the sparrow varies in different individuals, and often changes with the season. A single bird has been observed through several successive summers to sing nine or ten different sets of notes, usually uttering them one after another in the same order over and over. Careful attention will show almost any of our songsters to vary their melodies from time to time, but none have greater individuality than our subject. "Last season," writes John Burroughs, "the whole summer through, one sung about my grounds like this: *swee-e-t, swee-e-t, sweet, bitter*. Day after day, from May to September, I heard this strain, which I thought a simple but very profound summing up of life, and wondered where the little bird had learned it so quickly. The present season I heard another with a song equally original, but not so easily worded. Among a large troop of them in April, my attention was attracted to one that was a master songster—some Shelley or Thomson among its kind. The strain was remarkably prolonged, intricate, and animated, and far surpassed anything I ever before heard from that source."

Occasionally the song-sparrow sings on the wing while dropping to the ground from the top of a high tree—a favorite perch in early spring; and during the mating season many strange modifications of his tune strike the ear. As the summer comes on, his song, in common with that of all other birds, is less often repeated until it almost ceases in the fall; yet it may be heard, by an observing listener, every month in the year. His call to his mate is a simple *chuck* or *hwit*.

Rarely leaving his native copses until late in autumn, he has little need to exert large powers of flight, and moves from one low bush to another with a jerking, undulatory motion. His home is near the ground, and it is only the excitement of love which in spring prompts the males to seek the tree-tops.

His food is principally procured from the ground and among the branches and leaves of the wild shrubbery, and consists of blossoms, seeds, berries, and insects, varying according to the season and the age of its nestlings. Early in spring he is, as Mr. Gentry puts it, "a vegetarian," living

THE SONG-SPARROW. 175

upon the blossoms of the red maple and other early-blooming forest-trees, green ginger-berries, and the seeds of vegetables, in search of which it frequents the kitchen-gardens, and associates with the noble fox-sparrows and chattering goldfinches. As warm weather advances, the song-sparrow leaves the gardens, and seeks, in wilder spots, less of vegetable and more of animal food — eating strawberries, wild cherries, raspberries, etc., now and then as a relish; but depending for regular fare upon the young of the insect world just hatching out. It would be quite impossible to enumerate all the kinds eaten; probably everything palatable is welcome. I remember one June day watching one little fellow industriously picking very minute lice-like bugs from the under-side of the leaves of an apple-tree. He seemed inordinately fond of them, and swallowed twenty or thirty a minute, uttering the while a quick metallic chirp. Many kinds of caterpillars he likewise devours, among them clothes' moths and the loathsome tent-caterpillar, that stretches its canopied webs among the twigs of our orchard and shade trees, and drops down upon our heads in all its ugly nastiness; also ants, earthworms, and young beetles.

When the insects mature, and betake themselves beyond his easy reach, small fruits still remain; and, as these gradually disappear, he gives himself up more and more to a strictly graminivorous diet, breaking open the seed-vessels stored up by the wilderness of weeds growing in every field which the farmer has let "run to waste" for himself, but only thus cultivated the more for the sparrows. There is always enough of this material, either in the unbroken pods or fallen to the ground, to last through the winter such adventurous birds as brave our snows, screening themselves from the chilling blast in recesses of the dense thickets, or taking shelter under piles of logs and brush.

During the latter part of April, in ordinary seasons, the song-sparrow finds himself married, and he and his wife begin to construct their home. The site chosen is the green bank of some meadow brook, a tussock beside a country road, a hollow under some decaying log, where the nest shall be well secreted in a little thicket of grasses and flowers, or, in many cases, on bushes, vines, or even, as Mr. J. S. Howland assures me, in an old broken woodpecker's hole in an apple-tree. A friend in Astoria, Long Island, on May 8th, 1877, found a pair of these sparrows snugly ensconced in an ivy growing along the inner wall of a greenhouse. The birds had evidently watched their opportunity when the door was open or the glass raised during the warm days, and constructed their nest and deposited three eggs before they were discovered. In 1875 they built a nest in the same place, and the year before on the ground against the wall just outside. A pair has been around there for a great while; a nest being found within a hundred feet of the spot for some six or seven years. Wherever placed, it is a model and poetic bird-dwelling.

"What care the bird has taken not to disturb one straw or spear of grass, or thread of moss! You cannot approach it and put your hand into it without violating the place more or less, and yet the little architect has wrought day after day, and left no marks. There has been an excavation, and yet no grain of earth appears to have been moved. If the nest had slowly and silently grown, like the grass and the moss, it could not have been more nicely adjusted to its place and surroundings. There is absolutely nothing to tell the eye it is there. Generally a few spears of dry grass fall down from the turf above, and form a slight screen before it. How commonly and coarsely it begins, blending with the *débris* that lies about, and how it refines and comes to the centre, which is modelled so perfectly, and lined so softly."

Grasses are the timbers of the house—coarse stalks upon the outside, fine stems and soft leaves twined within; the edge of the nest overcast. It seems to be well proved that the nests found on the ground are built by young birds, while older and more experienced sparrows place their houses in vines and small trees, finding that at a little height they are less liable to danger; furthermore, these nests built at an elevation, being more exposed to the wind and less braced, are more compactly and skilfully constructed than those on the ground, the projecting ends of the straws being neatly interwoven, or tied down, so as to present a tolerably smooth exterior. The nests in the tussocks seem manufactured chiefly out of the dead stems of crab-grasses and other stuff within easy reach; but a variety of substances enter into the composition of the elevated nests, such as flowering weeds, narrow leaves, paper, strips of bark, and raw cotton (which sometimes thatches the whole outside), with horse-hair and milk-weed silk to give additional softness to the lining. When circumstances favor, a sort of sheltering platform is arranged over the nest in the tree or vines; just as frequently the approach to the nest hidden in the meadow lies through a tunnel like a field-mouse's path under the tall grasses.

The labor of building occupies the attention of the pair during the cool of the mornings and evenings of four or five busy days. Both birds work diligently, the male bringing the materials, and the female adjusting them.

The day after the nest is done an egg is laid, and one more each succeeding day until there are five; and very hard to distinguish from the eggs of several other ground-building sparrows they are. The ground-color runs through all intermediate tints from grayish or brownish white to decided green. The blotching is generally profuse, and often confluent into a wreath about the large end, the colors being underlying purples and bright brown surface painting. They are inclined to be thick and blunt rather than elongated, and will average about .90 by .60 of an inch. I can find no variations worth stating between the eggs of the different varieties. Those from the Pacific coast appear to be the largest, and those from southern localities the smallest; but the variety in size, shape, ground-color, and pattern is almost limitless, and I repeat that the strongest identification is necessary to make sure between these eggs and those of the swamp-sparrow, the grass-finch, the *Zonotrichiæ*, and several other members of the family.

The female sits eleven or twelve days, occasionally relieved by the male while she takes a brief rest. He assiduously provides her with food from hour to hour, but spends all his leisure at home, ready to resist invasion or insult, and enlivening the tedium of her sitting with his love ditties.

When the young are born, both parents are exceedingly devoted to their wants, carefully removing every trace of the old egg-shells and all foul matter far from the nest, and working with great energy to keep the hungry mouths filled. The nestlings are fed upon the young of many small insects, and as they grow older are given larger larvæ—earth-worms, house-flies, plant-lice, ants, and small night-flying moths. When twelve or thirteen days old, the young birds leave the nest, and in ten days more have learned to care for themselves. Meanwhile the mother has abandoned them to the father's guidance, and busies herself in the construction of a new home for a second family. Although left strong and neat, the first nest rarely seems to be used again; but the new one is built in close proximity to it. As before, the male is dutiful and loving, and the second brood is brought out in July, or sometimes earlier, so that even a third brood can be raised. But accidents or climate generally prevent this degree of success.

In autumn the song-sparrows are to be seen dodging about stone walls, roadside thickets, and old pastures, in little family companies of six or eight, no doubt consisting of parents with their second brood of young, which remain together in happy idleness, and move southward at their leisure.

Here the younger sons appear to have an advantage over their elder brethren of the first brood, who are early sent out to seek their fortunes, in that they enjoy the continued example and counsel of their parents during many weeks after they may be said to have "come of age," although possibly they may chafe under the restraints of paternal guidance, not to say old-fogyism, from which the youngsters of the first brood are now gayly delivered; but it would not be wonderful if it could be shown that the next year this latter brood, profiting by distasteful discipline, excelled in nest-building and in general prosperity over the others, who had enjoyed less advantages in the way of home education. Here is a new factor in the problem of natural selection.

VIII.

CIVILIZING INFLUENCES.

To say that the settlement of North America has produced a marked effect upon the animal life of the continent, and upon the birds as a part of the fauna, may seem too much of a truism to be worth discussion. Yet the degree to which this effect has been felt, and the various ways in which man's influence has been exerted upon animals, may still be objects of interesting inquiry. I confine myself alone to the effects produced by the white man, because the Indian seems to have caused hardly an appreciable change, either for good or evil, in the comparative plenitude, or in the habits, of the creatures dwelling about him. He himself was really as wild and indigenous as they, hunting, like the carnivores, purely for food, and, with the osprey, fishing only when his wants were urgent; his mind was too grim to entertain the idea of pursuing animals for sport, and his civilization too limited to cause much disturbance of natural conditions.

CIVILIZING INFLUENCES. 183

During the last two and a half centuries white men have spread everywhere, and in almost every part of the continent their machinery has replaced the original simplicity of nature. Thousands of square miles of forest have been cleared, marshes have been drained, rivers obstructed and tormented with mill-wheels, and cities have sprung up as swiftly as the second growth of scrub pines follows the levelling of an oak wood.

The inevitable result must follow that all our animals, birds included, would have been so harassed by their changed surroundings and the persecutions of human foes, that they would have rapidly disappeared. With the vast majority of the quadrupeds this has actually been the case. "Wild beasts" no longer haunt our forests, to the terror of the traveller; nor can the hunter now find game that a few decades ago was abundant almost at his door. It has been much the same with wildfowl and game birds. They have deserted their ancient nesting-places within our borders for the safer Arctic heaths, or old and young have been all but exterminated by gun and snare. Nevertheless, a large number of the smaller birds of our woodlands and prairies, as I hope to show, have been decidedly benefited by the advent of white men. I know of but one sort of quadruped—field-mice—of which this can also be said.

It is commonly observed that scarcely any small birds are seen in the depths of a forest, but they become abundant as one approaches the neighborhood of settlements. Travellers through Siberia know that they are coming near a village when they begin to hear the voices of birds, which are absent from the intervening solitudes. Every ornithologist has proved these facts in his own experience, and explorers who go to uninhabited and primeval regions have learned not to expect there the chorus that greets their ears from the great army of songsters thronging the fields in populous countries.

The song-birds — the small denizens of our summer groves, pastures, and meadows — seem, then, to recognize the presence of man's civilization as a blessing, and have taken advantage of it, both from love of human society and for more solid and prosaic reasons.

The settlement of a country implies the felling of forests, the letting in upon the ground of light and warmth, the propagation of seed-bearing cereals, weeds, and grasses enormously in excess of a natural state of things, the destruction of noxious quadrupeds and reptiles, and the introduction of horses and cattle. Each of these alterations of nature (except in some few cases, like that of the relation of the woodpecker to the cutting away of timber) is a direct benefit to the little birds. It is not difficult to demonstrate this.

Birds naturally choose sunny spots in which to build their nests, such as some little glade on the bank of a stream; when roads were cut and fields levelled in the midst of sombre woods, the area suitable for nesting was of course greatly added to, and a better chance thus afforded for successfully hatching and rearing broods of young. The way in which the wood-roads cut by the hemlock bark-peelers through the dense forests that clothe the remote Catskills have become the haunt of birds and insects, is a capital example in urging this point. One of the largest avian families—that of the sparrows, finches, and buntings —subsists almost exclusively on seeds of weeds and grasses; and the members of a large proportion of other families depend somewhat for their daily supply on this sort of food. Under the universal shade of trees weeds can grow only sparingly, and on prairies the crop is often killed by drought, or is burnt in the autumn; but the cultivation of immense fields of grain and hay, and the making of broad pastures and half-worn roads, which almost immediately become filled with weeds, has furnished the birds with an inexhaustible and unfailing harvest.

Birds suffer much harm from many quadrupeds—foxes,

weasels, skunks, rats, etc.—which catch them on their roosts, suck their eggs, and kill their fledglings. Snakes also are fond of them, and destroy many nests every season—in early summer subsisting almost alone on eggs. All these animals, particularly foxes, skunks, and serpents, are greatly reduced in number by settlements, although it must be confessed that their absence is somewhat compensated for by the introduction of domestic cats, which go foraging through the woods, to the grief of all their feathered inhabitants. No longer in fear of their natural enemies, and learning that there is little reason to be apprehensive of harm from mankind, the small birds forsake their silent, shy manners, come out of the thickets where they have been hiding, and let their voices be heard in ringing tones, easily interpreted as rejoicing at deliverance from fear, and thanksgiving for liberty to sing as loud as pleases them.

All small birds are more or less completely insectivorous (even the cone-billed seed-eaters having to feed their young with larvæ at first), and naturally congregate where this food is most abundantly supplied. There would seem to be enough anywhere; but the ploughing and manuring of the soil facilitates the growth and increase of such insects as go through their metamorphoses in the ground; and the culture of orchards furnishes an excellent resort for many boring and fruit-loving moths, beetles, and the like, which find the best possible circumstances for their multiplication in the diseased trunks and juicy fruit of the apple, plum, cherry, and peach. No part of the farm has so many winged citizens as the orchard.

The presence of horses, cattle, and sheep offers to flies and other insect tribes excellent opportunities for the safe rearing of their eggs in the dunghills and heaps of wet straw always lying about barns, and attracts a great colony of those minute beetles upon which the fly-catching birds principally maintain themselves. The cattle-yard, therefore, forms a sort of game-preserve for such birds, and many species flock thither. Swallows are hardly ever found except in the vicinity of barns; the cow-bunting receives its name from its habit of constantly associating with cattle; and the king-bird finds the stable-yard his most profitable hunting-ground. Near the habitations of men, small birds also enjoy protection from hawks and owls, which hesitate to venture away from the shelter of the woods, and whose numbers are reduced, unwisely perhaps, by incessant persecution.*

---

* In several States of the Union bounties are offered, sometimes by county authorities, sometimes by game-protective associations, and hundreds of hawks and owls are killed annually.

The logic of the case is simple; birds will assemble chiefly where food for themselves and their young is in greatest abundance, and where they are least exposed to enemies. These two prime conditions of prosperity, with many favorable concomitants, man's art supplies to the insessorial birds, which, on the other hand, suffer little direct injury from his contact. Yet some species seem little affected by the civilizing of the country, either in numbers or habits, while others increase rapidly on the first settlement of a region, and then decrease again. Of this class are the prairie-hen (*Cupidonia cupido*) and the mallard. "They find abundance of food in the corn and wheat fields; while the population is sparse and larger game so abundant, they are hunted very little; but as the population increases they are gradually thinned out, and become in some cases exterminated. Other birds, as the quail, are wholly unknown beyond the frontier, and only appear after the country has been settled a short time. Still others, woodland species, appear in regions where they were never known before, as groves of trees are planted, and thick woods spring up on the prairies as soon as the ravages of the fires are checked."

Striking examples of how some of our birds have accepted this tacit invitation to make men their confidants occur in the history of the American swallows and swifts. Our purple martins spread themselves in summer all over North America, but are becoming rare in the New England States, whence they seem to have been driven by the white-bellied swallows, which have gradually grown more numerous, and which, preceding the martins in the spring, take possession of all the boxes put up for the accommodation of the martins, and exclude the rightful tenants *vi et armis*. Their natural nesting-places were hollow trees and cavities in rocks; but now, throughout the whole breadth of the land, it is rare to find martins resorting to such quarters, except in the most remote parts of the Rocky Mountains. They have everywhere abandoned the woods, and come into the villages, towns, and even cities, choosing to nest in communities about the eaves of houses and barns, and in sheltered portions of piazzas, or to take possession of garden birdboxes, where their social, confiding dispositions have rendered them general favorites.

A very similar case is presented in the case of our chimney-swift, which finds a chimney a far more desirable residence than a hollow tree in the woods.

Other species of American swallows afford still more striking illustrations of a change in the manner of life effected by association with men. Perhaps the most curious example is the case of the cave-swallow (*Petrochelidon lunifrons*). This bird remained undiscovered until 1820, when it was met with by the celebrated Thomas Say when naturalist to Major Long's expedition to the Rocky Mountains, a memento of which remains in the name of one of the loftiest heights of the snowy range—Long's Peak. In 1825, however, the bird suddenly appeared at Fort Chippewa, in the Fur Country, and contentedly built its nest under the eaves. Even earlier it had been seen on the Ohio River, at Whitehall, New York, and very soon after was found breeding in the Green Mountains, in Maine, in New Brunswick, and among the high limestone cliffs of the islands along that precipitous coast. It occurs also westward to the Pacific coast. It is hardly to be supposed that these swallows were indigenous to some restricted locality in the West, whence they suddenly made such a startling exodus; but rather it is believed that they always had existed in isolated spots all over the country, but so far apart, and so uncommonly, that they were overlooked.

The experience of the barn-swallow (*Hirundo horreorum*) has been much the same; and the Rocky Mountain swallow (*Tachycineta thalassina*), which breeds in far-separated colonies throughout the mountainous West, is fast following its example in scraping acquaintance with mankind.

The natural breeding-place of all the three species I have mentioned is in caves and crevices of rock, the irregularities and hollows of limestone crags affording them the best chances. "Swallows' Cave," at Nahant, is remembered as one of their hospices. I have seen all three species breeding together among the ragged ledges of Middle Park, Colorado; but considerable differences were noticeable between the houses of these uncivilized builders and those of their educated brethren at the East, who now, perhaps, would find it rather hard to rough it as did their ancestors.

Under the shelter of warm barns, and with such an abundance of food at hand that they have plenty of leisure between meals to cultivate their tastes and give scope to their ingenuity, our barn and eave swallows have shown a wondrous improvement in architecture. The nests of the barn-swallows that I saw at the hot sulphur springs in Colorado consisted only of a loose bed of straw and feathers, for the hollow floors of the niches in which they were placed formed cavity and barrier for the safety of the eggs. Some nests, resting on more exposed ledges, had a rude foundation and rim of mud, but did not compare with the elaborate half-bowls, lined with hay and feathers, that are plastered by the same species so firmly against the rafters of our barns, or with the large nest that is balanced on the beam, with its edges built up so high that the callow young can hardly climb, much less tumble out, until quite ready to fly. Nevertheless, the general character of the nest is the same; the eastern, civilized swallows have only made use of their superior advantages to perfect the inherited idea. In the case of the barn-swallow, its civilization results in an addition to its pains (is it not a natural consequence?), in that its nest now is required to be much larger, more carefully, and hence more laboriously, made. On the other hand, its neighbor, the eave-swallow, has contrived to save itself labor by the change from wild life.

This latter species is sometimes called the republican swallow, because at the breeding-season it gathers in extensive colonies, where its homes are crowded together as closely as the cells in a honey-comb, one wall often serving for two or more contiguous structures. The nests are gourd-shaped, or like a chemist's retort, and are fastened by the bulb to the cliff, generally where it overhangs, with the curving necks opening outward and affording an entrance just large enough to admit the owner. This retort is constructed of pellets of mud, well compacted in the little mason's beak, and made adhesive by mixture with the glue-like saliva with which all swallows are provided. In this snug receptacle the pretty eggs are laid upon a bed of soft straw and feathers. Such was the elaborate structure deemed necessary by the swallows so long as they nested in exposed places, where they had to guard against the weather and crafty enemies. "But since these birds have placed themselves under the protection of man, they have found that there is no longer any need of all this superfluous architecture, and the shape of their nest has been gradually simplified and improved. In 1857, on one of the islands in the Bay of Fundy, Dr. T. M. Brewer met with a large colony whose nests, on the side of the barn, were placed between two projecting boards put up for them by the friendly proprietor. The very first year they occupied these convenient quarters, every one of these sensible swallows built nests open at the top, discarding the old patriarchal domes and narrow entrances of their forefathers." This is not an isolated case, but rather has come to be the rule wherever there was a roof over them, so far as my own observation goes.

The purple martin and white-bellied swallow both accept of houses ready made, saving themselves all trouble except in furnishing them; and even the burrowing-bank and rough-winged swallows are learning that it is cheaper to build in a snug cranny in an old wall than laboriously to dig a deep crypt in a sand-bank wherein to lay their pearly eggs.

Men's industries have supplied the birds with some new and exceedingly useful building materials, such as furnishing those weavers, the orioles and vireos, with strings and yarn for the warp of their fabrications, and the yellow-bird with cotton and wool to make her already downy bed still softer. Instances of abnormally late and early breeding seem to be very common in England, and are coming to be more and more frequently recorded on this side of the Atlantic. This is not to be wondered at, since our operations insure to the birds a continued supply of suitable food, and thus enable them to rear their young at seasons when in a wild state it would not be possible to do so. The English sparrows, breeding all the year round, or nearly so, in the parks of our coast cities, are a case in point.

That civilization has to some extent governed the migrations and geographical distribution of many species of our birds not directly warred upon as pot game, for amusement, or because they are obnoxious to crops, could easily be shown had I space allowed me to bring forward illustrations; and when another two centuries have rolled around the effect will be very striking. The mocking and Bewick's wrens, the rose-breasted grossbeak, chestnut-sided warbler, and other species, have spread northward and become more abundant since the time of Wilson and Audu bon; the bobolink has kept pace with the widening cultivation of rice and grain fields; the red-headed woodpecker has retreated from New England; the Arkansas flycatcher has multiplied and spread as a town bird through all the cities and villages from Council Bluffs to Denver; the raven has gradually retired before the wood-cutter, until it has almost ceased to exist; while year by year the crow has extended its range, without seeming in the least to diminish its force in the older districts, but crowding the wild and refractory raven farther and farther beyond the frontier.

Although none have abandoned their old way of life so completely as the swallows, many other birds have profited by the constructions and friendship of the human race. The bluebird and house-wren, chickadees and nuthatches dig holes in the fence-posts conveniently rotting for their use; and even such wild species as the western flycatcher, great-crested kingbird, and Bewick's wren, occasionally attach themselves to mankind, and hatch their young under his roof for greater security. Even the whippoorwill and nighthawk, asleep all day in the swamp, are glad to come to the farmer's house in the evening, and now and then to deposit their eggs on a flat roof. In the Rocky Mountains I have seen flocks of white ptarmigans nimbly hopping around the door-steps of miners who were seeking silver far above timber-line, picking up the crumbs thrown to them, as tame as pet chickens.

In not a few instances, here as well as abroad, superstition brings profit to our birds. An honest old Pennsylvania Dutchman, around whose barn clouds of swallows hovered, told Wilson that he must on no account shoot any, for if one was killed his cows would give bloody milk, and that so long as the swallows inhabited the barns his buildings were in no danger of being struck by lightning. The arrival of the fish-hawk or osprey on the New Jersey coast, at the vernal equinox, notes the beginning of the fishing-season. In some parts of New England the appearance of the golden-winged woodpecker means the same thing, for the bird is known as the "shad-spirit." The coming of both is therefore hailed with satisfaction, and it is considered so "lucky" to have an osprey nesting upon one's farm, that proprietors cherish its huge house in the lone tree with uncommon care, recalling the reverent fostering that a family of storks will enjoy from the peasant of the Netherlands on whose roof their nest has been placed.

The result of all these circumstances, as it seems to me, is, that the aggregate army of singing birds in the United States, east of the Mississippi, has been very considerably enlarged during the last two centuries, and is still on the increase. This can be owing only to the fact that by cutting down the forests, etc., civilized man has tempered the rigor of the climate, has multiplied the sources of bird food, has appended many additional places suitable for the young, and has enabled more fledglings to be brought to maturity by reducing the ranks of the enemies of the birds. This has not only augmented their number, and very appreciably modified their habits of nesting and migration, but probably has somewhat changed even their physical and mental characteristics. There is little doubt in my mind, for instance, that in making their lives less laborious, apprehensive, and solitary, man has left the birds time and opportunity for far more singing than their hard-worked, scantily-fed, and timorous ancestors ever enjoyed—a privilege a bird would not be slow to avail itself of.

But, on the other hand, it seems to me equally certain that the music of our more domestic birds, though greater in volume, is not so sweet in tone as that of their wilder brethren. Our foreign street sparrows (the London " Jims ") are naturally, I suppose, rather harsh-voiced; but, whatever they might have been a thousand years ago, they could hardly be otherwise now, when the rattle-te-bang of the city pavements has been their only teacher for many centuries. The mocking-bird has learned to imitate the screech of the ungreased wheelbarrow and the howl of the farmer's dog—no dulcet sounds. Many of the noises constantly uttered by men and evoked by their work are anything but melodious, and young birds born and bred in their midst must surely turn out less sweet and accomplished singers than if reared among the gentle whisperings of leafy woods, and learning music only from the golden-mouthed minstrels of the sylvan choir.

IX.

HOW ANIMALS GET HOME.

One of the most striking powers possessed by animals is that of finding their way home from a great distance, and over a road with which they are supposed to be unacquainted. It has long been a question whether we are to attribute these remarkable performances to a purely intuitive perception by the animal of the direction and the practicable route to his home, or whether they are the results of a conscious study of the situation, and a definite carrying out of well-judged plans.

Probably the most prominent example of this wonderful power is the case of homing pigeons. These pigeons are very strong of wing, and their intelligence is cultivated to a high degree; for their peculiar "gift" has been made use of since "time whereof the memory of man runneth not to the contrary." The principle of heredity, therefore, now acts with much force; nevertheless, each young bird must be subjected to severe training in order to fit it for those arduous competitions which annually take place among first-rate birds. As soon as the fledgling is fairly strong on its wings, it is taken a few miles from the cot and released. It rises into the air, looks about it and starts straight away for home. There is no mystery about this at all; when it has attained the height of a few yards the bird can see its cot, and full of that strong love of home which is so characteristic of its wild ancestors, the blue-rocks, it hastens back to the society of its mates. The next day the trial-distance is doubled, and the third day is still further increased, until in a few weeks it will return from a distance of seventy miles, which is all that a bird-of-the-year is "fit" to do; and when two years old, will return from two hundred miles, longer distances being left to more mature birds. But all this training must be in a continuous direction; if the first lesson was toward the east, subsequent lessons must also be; nor can the added distance each time exceed a certain limit, for then, after trying this way and that, and failing to recognize any landmark, the bird will simply come back to where it was thrown up. Moreover, it must always be clear weather. Homing pigeons will make no attempt to start in a fog, or if they do get away, a hundred chances to one they will be lost. Nor do they travel at night, but settle down at dusk and renew their journey in the morning. When snow disguises the landscape, also, many pigeons go astray. None of these circumstances seriously hampers the semi-annual migrations of swallows or geese. They journey at night, as well as by day, straight over vast bodies of water and flat deserts, true to the north or south. Homing pigeons fly northward or southward, east or west, equally well, and it is evident that their course is guided only by observation. Watch one tossed. On strong pinions it mounts straight up into the air a hundred feet. Then it begins to sweep around in great circles, rising higher and higher, until—if the locality is seventy-five or one hundred miles beyond where it has ever been before—it will go almost out of sight. Then suddenly you will see it strike off upon a straight course, and that course is homeward. But take the same bird there a second time and none of these aërial revolutions will occur—its time is too pressing, its homesickness too intense for that; instantly it turns its face toward its owner's dove-cot.

These facts mean something. They show that two definite intellectual processes serve to decide for the bird the direction he is to take—observation and memory. He gets high enough, and turns about times enough, to catch sight of some familiar object, and he makes for it; arrived there, another known feature catches his eye, and thus by ever narrowing stages he is guided home. Few persons have any idea of the distance one can see at great elevations. More than once I have stood on the Rocky Mountains, where, had I been a pigeon, I could have steered my flight by another mountain more than one hundred miles distant. Balloonists say that at the height of half a mile the whole course of the Thames or the Seine, from end to end, is spread out as plain as a map beneath their eyes. There is no doubt that a pigeon may rise to where he can recognize, in clear weather, a landscape one hundred and fifty miles away; it has been done repeatedly, though only by the best birds, specially trained for that particular line of flight. There is no greater error than to suppose that carrier-pigeons sent a long distance from home in any direction will always return, as though attracted by a loadstone. The benevolent lady received only a good-natured laugh for her pains, when she offered to equip the late British Arctic expedition with these winged messengers, who, she supposed, could be despatched from any point with tidings, and have a fair chance of getting straight back to England.

A pigeon's power of memory is really wonderful. Beginning with short stages, perhaps of not more than a dozen miles, the final stage of a match-flight of five hundred miles will be more than one hundred. The country has been seen but once, yet the bird remembers it, and not only for the three or four days of a match, but for months. In June, 1877, birds trained from Bath to London were twice flown. On June 11th of 1878 they repeated the trip at good speed. Such feats are not uncommon with Belgian birds—the best of all—and there have been several authenticated instances of their going off-handed from England to Belgium after having been kept in confinement many months. But the homing intelligence of pigeons is subject to much irregularity of action, and this very circumstance insists that it shall not be considered an unvarying, unreasoning instinct.

Enough has now been said, perhaps, to enable one to see that, however much the bird may be aided by an acute sense of direction—a capability, I mean, of preserving a straight course, once ascertained, which sense some may prefer to speak of as an "instinct"—the homing faculty of *le voyageur pigeon* is the result of education, and is not a matter of intuition at all.

The bee pursues a truly similar course. When he is loaded with nectar, you will note him cease humming about the heads of the flowers and spring up in a swift, vertical spiral, and after circling about a moment, shoot homeward "in a bee-line." Evidently he has "got his bearings." Had you watched him the first time he ever left his hive you would have observed precisely similar conduct to acquaint himself with the surroundings.

How a bird like the albatross, the man-of-war-hawk, or the petrel, swinging on tireless pinions in apparently aimless flight over the tossing and objectless ocean, suddenly rouses its reserve of strength to traverse in a day or two the hundreds of miles between it and the rocky shores where it builds its nest; or how it finds the lone islet which these winged wanderers of the sea alone render populous, is not easily explained. Nor can we readily understand how once a year the salmon comes back (from conjecture only guesses where), not to the coast alone, for that would be no more than an ordinary case of migration, but to the identical stream where it was born; and to prove that it was not a blind emotion that led it, would be harder than in the case of the pigeon, the bee, or even the frigate-bird. Yet who knows that the fishes may not be able to perceive the differences in the water which we designate "variations of temperature and density," or still more delicate properties, and thus distinguish the fluid of their native place from the outside element? It is a quetion, however, whether this phenomenon comes properly within the scope of this article.

Many domestic animals show a true homing faculty, and often in a degree which excites our surprise. One of the most remarkable cases I knew was that of two of the mules of a pack-train which, plainly by concerted action, left our camp one morning without cause or provocation. We were in south-western Wyoming, about seventy-five miles north-west of Rawlins Station, where we had begun our march. Our course, however, had been an exceedingly roundabout one, including a great deal of very bad country, where no road or trail existed. These mules made no attempt to trace it back, but struck straight across the country. They were chased many miles, and showed not the least hesitancy in choosing their way, keeping straight on across the rolling plain, with a haste which seems not to have been diminished until Rawlins was almost reached, when they were caught by some prospectors. For weeks they had to be kept carefully hobbled to prevent a repetition of the experiment.

How did these animals know the direction with such certainty? Mules frequently follow a very obscure trail backward for many miles, and, even more than horses, may be trusted to find the way home in the dark; but this is only when they have been over the road before, and is quite as fully due to their superior eyesight as to their strong sense of locality. I have also seen mules following the trail of a pack-train a few hours in advance, almost wholly by scenting; but the two runaways before mentioned had no other conceivable help in laying their course than some distant mountain-tops north and east of (and hence behind) them, and to profit by these would have required a sort of mental triangulation.

But the most common instances of homing ability are presented by our domestic pets, which often come back to us when we have parted with them, in a way quite unaccountable at first thought. An extremely instructive series of authentic examples of this were published in successive numbers of that excellent newspaper, the *London Field*. The discussion was begun by a somewhat aggressive article by Mr. Tegetmeier, in which he expressed the opinion that most of such stories current were "nonsense," and cordially assigned to the regions of the fabulous those narratives which seemed to attribute this power to a special faculty possessed by the animal, instancing himself two cases where a dog and a cat found their way home, as he very justly supposes, by using their memories. The distance was not great; they obtained a knowledge of the routes, and took their departure. "Very interesting," replied a correspond ent, " but no argument against another cat or dog home-returning twenty or thirty miles across a strange district by means of instinct." And as evidence of his conclusion that " there is an attribute of animals, neither scent, sight, nor memory, which enables them to perform the home-returning journeys," this gentleman said:

"When I resided at Selhurst, on the Brighton and South Coast Railway, a friend living at Sutton gave me an Irish retriever bitch. She came over to him about a month previously from the County Limerick, where she was bred; and during her stay at Sutton she was on chain the whole time, with the exception of two walks my friend gave her in the direction of Cheam, which is in an opposite quarter to Selhurst from Sutton. She came to me per rail in a covered van, and the distance from home to home is about nine miles. She was out for exercise next morning, ran away, and turned up at her previous home the same afternoon."

But this proved to be a mild instance of such performances. A fox-hound was taken by train in a covered van forty miles from the kennels of one hunt to those of another in Ireland. The hound was tied up for a week, and then she was taken out with the pack. She hunted with them for the day, and returned in the evening to within a hundred yards of the kennel. "Here," relates the narrator, "I noticed her go into a field, sit down, and look about her. I called out to the young gentleman who hunts the hounds, whose way home was the same as mine: 'J., Precious is not going on with you.' 'Oh, there's no fear of her,' was the reply. 'As she came so far, she will come the rest of the way.' So we went on to the kennel close by, but Precious did not appear, and we came back at once to the spot, sounded the horn, and searched everywhere. That was at six o'clock in the evening. On the following morning at six o'clock, when the messman went to the kennel door at Doneraile, Precious was there."

An officer took a pointer which certainly had never been in Ireland before, direct from Liverpool to Belfast, where he was kept for six months at the barracks. He was then sent by train and cart, in a dog-box, thirty-four miles into the country, and tied up for three days. Being let out on the morning of the fourth, he at once ran away, and was found that same evening at the barracks at Belfast.

A sheep-dog was sent by rail and express wagon from the city of Birmingham to Wolverton, but, escaping from confinement the next Saturday at noon, on Sunday morning reappeared in Birmingham, having travelled sixty miles in twenty-four hours. Says one writer: "I was stopping with a friend about eighteen miles from Orange, New South Wales. My host brought a half-grown kitten sixteen miles by a cross-bush track, tied in a flour-bag at the bottom of a buggy. She was fed that night; in the morning she had disappeared. She was home again in rather less than four days." The same person owned a horse in the interior of Australia, which, after two years of quiet residence on his run, suddenly departed, and was next heard of one hundred miles away, at the run of the old master from whom it had been stolen years before.

A rough-coated cur was taken by a gentleman to whom he had been given from Manchester to Liverpool by train, thence to Bangor, North Wales, by steamboat; but on landing at Bangor the dog ran away, and the fourth day afterward, fatigued and foot-sore, was back in his home kennel, having undoubtedly travelled straight overland the whole distance. The same gentleman knew of a kitten that was carried in a covered basket six miles from one side of Manchester to the other, and found its way back the next day through the turbulent streets. Similarly, a fox-hound transported in a close box between points one hundred and fifty miles distant, and part of the way through the city of London, came back as soon as let loose. A retriever bitch did the same thing from Huddersfield to Stroud, a fortnight after being taken to the former place by rail; and a fox-hound returned from Kent to Northamptonshire, which are on opposite sides of the Thames; finally a dog came back to Liverpool from a distant point, whither he had been forwarded by rail *in the night*.

So many such instances are recorded that I refrain from mentioning more, except a couple of very illustrative ones which I find vouched for in the Rev. J. G. Wood's valuable little book, "Man and Beast." A mechanic who worked in Manchester, but lived at Holywell, Wales, having been home on a visit, was given a dog to take back with him. "He led the animal from Holywell to Bagill by road, a distance of about two miles. Thence he took the market-boat to Chester, a distance of about twelve miles, if I remember right. Then he walked through Chester, and took rail for Birkenhead. From that station he walked to the landing-stage, and crossed the Mersey to Liverpool. He then walked through Liverpool to the station in Lime Street. Then he took rail to Manchester, and then had to walk a distance of a mile and a half to his home. This was on Wednesday. He tied the dog up, and went to his work on Thursday as usual; and on the Sunday following, thinking that the dog was accustomed to the place, he set it at liberty. He soon lost sight of it, and on the Wednesday following he received a letter from his mother, stating that the dog had returned to her. Now you will see that the dog went first by road, then by market-boat, then through streets, then by rail, then by steamer, then through streets again, then by rail again, then through streets again, it being dark at the time." Whether the animal really did follow the backtrack with all this exactness or not, one thing is certain, he had sagacity enough to find his way, and (as is noteworthy in all these incidents) did so with astonishing speed.

The second instance is still more striking, and illustrates very forcibly the strong love of home in the dog, which is the motive in all these extraordinary and difficult journeys. "A gentleman in Calcutta wrote to a friend living near Inverkeithing, on the shores of the Frith of Forth, requesting him to send a good Scotch collie dog. This was done in due course, and the arrival of the dog was duly acknowledged. But the next mail brought accounts of the dog having disappeared, and that nothing could be seen or heard of him. Imagine the astonishment of the gentleman in Inverkeithing when, a few weeks later, friend Collie bounced into his house, wagging his tail, barking furiously, and exhibiting, as only a dog can, his great joy at finding his master." Inquiry showed that the dog had come aboard a Dundee collier from a ship hailing from Calcutta.

Comparing all these examples and many others—for hundreds, almost, of similar cases with various animals might be cited—certain general facts appear.

First, incidentally, brutes equally with men become homesick. Those that stay away, as well as those that return to their former homes, show this very plainly, and often pitiably. This feeling is the motive which leads them to undergo perils and hardships that no other emotion would prompt them to undertake or enable them to endure. But it is the *most thoroughly domesticated* and *most intelligent* breeds of animals that this homesickness attacks the most severely; while, correlatively, the most difficult feats of finding their way home are manifested by the same class. It is the finely-bred horses, the carefully-reared pigeons, the highly-educated pointers, fox-hounds, and collies that return from the longest distances and over the greatest obstacles.

This would seem to indicate that the homing ability is largely the result of education; whatever foundation there may have been in the wild brute, it has been fostered under civilizing influences, until it has developed to an astonishing degree. I would like to ask any one who believes that this ability is wholly a matter of intuition—an innate faculty—why such an instinct should have been planted in the breast of animals like dogs and horses in their wild condition? They had no homes to which they could become attached as they do now in their artificial life; or when they did settle during the breeding season in any one spot, either they did not quit it at all, wandered only for a short distance, or else the females alone remained stationary, while the males roved as widely as usual. There would seem to be no call, therefore, for such an instinct in the wild animal. That they may always have had, and do now possess, a very acute sense of direction, enabling them to keep the points of the compass straight in their minds far better than we can, I am willing to admit; but I doubt whether the evidence proves a nearer approach to a homing "instinct" than this. On the contrary, I believe, as I have already hinted, that beyond this the performances of animals in the line of our inquiry are the result of accurate observation and very retentive memory. That all these animals now and then do miss their bearings, get "turned around" and wholly lost, is true, and is a fact to be remembered in this discussion.

In the case of the birds, observation by sight is sufficient. They rise to a height whence they can detect a landmark, and flying thither, catch sight of another. The experience of pigeon-trainers shows this satisfactorily, and that of the falconers supports it. The far-reaching eyesight of birds is well known. Kill a goat on the Andes, and in half an hour flocks of condors will be disputing over the remains, though when the shot was fired not a single sable wing blotted the vast blue arch. The same is true of the vultures of the Himalayas and elsewhere. Gulls drop unerringly upon a morsel of food in the surf, and hawks pounce from enormous heights upon insignificant mice crouching in fancied security among the meadow stubble, while an Arctic owl will perceive a hare upon the snow (scarcely more white than himself) three times as far as the keenest-eyed Chippewa who ever trapped along Hudson's Bay. The eyesight, then, of pigeons and falcons is amply powerful to show them the way in a country they have seen before, even though the points they are acquainted with be a hundred miles apart.

In the cases of horses, dogs, and cats the explanation may be more difficult, and not always possible to arrive at. Horses and mules are extremely observant animals, and quick to remember places; everybody who has ever had anything to do with them must know this. Their recollection is astonishing. The Rev. J. G. Wood tells of a horse which knew its old master after sixteen years, though he had grown from a boy to a man, and was, of course, much changed in both voice and appearance. It is probable that where horses come back, they do so mainly by sight and memory.

As for dogs, they not only can see well, but they have the additional help of their intelligent noses. The proficiency to which some breeds of dogs have brought their smelling powers—the precision with which they will analyze and detect different scents—is surprising. I have lately seen trustworthy accounts of two hunting-dogs, one of which pointed a partridge on the farther side of a stone wall, much to the surprise of his master, who thought his dog was an idiot; and the other similarly indicated a bird sitting in the midst of a decaying carcass, the effluvium of which was disgustingly strong, yet not sufficiently so to disguise the scent of the bird to the dog's delicate nostrils. Fox-hounds will trace for miles, at full speed and with heads high, the step of a Mercury-footed fox, simply by the faint odor with which his lightly touching pad has tainted the fallen leaves.

There are few cases where a dog is taken from one home to another, when he could not see most of the time where he was going. In that complicated journey of the Holywell workman's pet from northern Wales to Manchester, the little fellow had his eyes open the whole distance, we may be sure, and if he could speak he would no doubt tell us that he remembered his previous journey pretty well. But many times, especially where transported by rail, it is unquestionable that dogs rely upon their noses to get them back. Finding that they are being kidnapped, carried off from home and friends in this confined, alarming fashion, unable to see out of the tight box or the close car, they do just what you or I would under similar circumstances—exert every possible means left them of discovering whither they are going, and take as many notes as possible of the route, intending to escape at the very first opportunity. One means of investigation remaining is the scent, and this they would use to great advantage, examining the different smells as their journey progressed, and stowing them away in their memory to be followed back in inverse order when they have a chance to return. Granting to these animals the discriminating sense of smell which experience shows to be possessed by them, I do not see any reason why they should not be able to remember a journey by its succession of odors just as well as they would by its successive landmarks to the eye. Even we, with our comparatively useless noses, can smell the sea from afar; can scent the sweetness of the green fields as well as the smokiness of black towns; and can distinguish these general and continuous odors from special or concentrated odors, which latter would change direction as the smeller changed position. How far this sense has really been developed in the human subject, perhaps few know; but in the history of Julia Brace, the deaf and blind mute of Boston, for whom the late Doctor Howe accomplished so much, occurs a striking example. In her blindness and stillness, Julia's main occupation was the exercise of her remaining senses of touch, taste, and smell. It was upon the last, we are told, that she seemed most to rely to obtain a knowledge of what was going on around her, and she came finally to perceive odors utterly insensible to other persons. When she met a person whom she had met before, she instantly recognized him by the odor of his hand or glove. If it was a stranger, she smelled his hand, and the impression remained so strong that she could recognize him long after by again smelling his hand, or even his glove, if he had just taken it off; and if, of half a dozen strangers, each one should throw his glove into a hat, she would take one, smell it, then smell the hand of each person, and unerringly assign each glove to its owner. She would pick out the gloves of a brother and sister by the similarity of odor, but could not distinguish between them. Similar cases might be produced, though hardly one of superior education in this respect; and in the light of it, it is not difficult to suppose that a sharp dog should be able to follow back a train of odors that he had experienced shortly before.

But there is another way by which anxious animals may learn their route both going and coming, and that is by listening and inquiring. It is remarkable how much of what is said by their masters all dogs understand. The books and periodicals of natural history and sport abound with illustrations of this, and one lately occurred within my own experience. A very good-natured and amusing, but utterly unthoroughbred, little dog was a member of a family which I was visiting. The dog and I became very good friends at once, and remained so until the second day, when I casually began to joke his master upon owning such a miserable cur. At once the little dog pricked up his ears, and, noticing this, I continued my disparagements in a quiet, off-hand tone, his master meanwhile defending and condoling with him, until at last the dog could stand it no longer, but, without any provocation beyond my language, which was not addressed to him at all, sprung up and softly bit at my heel, as though to give me warning of what might happen if the joke went any further; and after that he utterly broke off our friendship.

I mention this incident to call attention to the alertness of our household pets in hearing and comprehending what is being said. Could not a dog on a railway remember the names of the towns through which he passed as they were called out by the attendants and spoken by travellers, and so be able to judge something of his way in return? The Rev. Mr. Wood suggested that the collie which returned from India was enabled to find the right vessel at Calcutta by hearing the well-known language and accent of the Scotch sailors; and again picked out from among many others the right collier in which to finish the journey, partly by remembrance of the rig, but also by recognizing the still more familiar and home-like dialect of the Dundee men. In a country where dialects are so marked as in Great Britain, this sort of observation would no doubt be of great help to an intelligent animal. Take the case of the Holywell workman's dog. It is quite possible that he discovered the right route from Liverpool, whither it would not be so difficult to make his way from Manchester, by following some rough-tongued Welshman until he found himself among his own hills again.

But there is still more to be said about this part of a homesick animal's resources and ingenuity. I am firm in my belief that animals have a language of signs and utterances by which they communicate with each other, and that their vocabulary, so to speak, is much larger than it has generally been considered to be. Dupont de Nemours declared that he understood fourteen words of the cat tongue. I am perfectly convinced that those two wicked little mules of ours, which ran away so disgracefully from our camp in Wyoming, had planned the whole thing out beforehand, and thus very likely had made up their minds as to the road. They had been bitter enemies, biting and kicking each other, contesting for coveted places in the line, and quarrelling the whole trip. But the evening before they ran away they were observed to be very amicable. It attracted our notice, and the last that was seen of them in the morning, just before they bolted, they stood apart from the rest with their heads together and their ears erect, waiting the right moment to dart away together. Tell a mountain mule-driver that the little beasts do not talk among themselves (chiefly in planning cunning mischief), and he will laugh in your face.

Cats, we know, consult a great deal together, and two street dogs often become great cronies. Why should not these dogs and cats be able to tell stray companions something which should help them on their way? I believe they do—just how, I don't pretend to say.

It seems to me, therefore, that the examples cited above, and a host of others like them, show that all domestic animals have a very strong love of places and persons. In many cases this homesickness is so strong as to lead them to desert a new abode, when transferred to it, and attempt to return to their former home; but they *rarely or never do so without having a definite idea in their minds as to the route*, although it is often very long and circuitous, and hence they almost invariably succeed; *otherwise, they do not try.* It is not every animal, by a long list, that deserts a new home the moment the chain is loosed; only one, now and then. In regard to the method used by them to find their way, it appears that they have no special instinct to guide them, but depend upon their memory of the route, the knowledge of which was acquired by an attentive study through the senses of sight, smell, and hearing, possibly by communication with other animals. The phenomenon, as a whole, affords another very striking example of animal intelligence.

X.

A MIDSUMMER PRINCE.

Cecilius Calvert, second Baron of Baltimore, has a hold upon the recollections of mankind far surpassing that secured by any monument in the noble city which he founded, in the fact that the most charming bird that makes its summer home in the parks of that city bears his name. That bird is the Baltimore oriole—*Icterus baltimore* of Linnæus. Its plumage is patterned in orange and black, the baronial colors of the noble lord's livery; and Linnæus only paid an appropriate compliment to the source to which he owed his specimen of the new species, when, in 1766, he recognized the coincidence in the name.

Then as now the orioles were among the most beautiful and conspicuous of woodland birds. From their winter retreat under the tropics they return northward as the warm weather advances, arriving in Maryland during the latter part of April, and reaching central New England by the middle of May. In these migrations, performed mostly by day, they fly continuously and in a straight line high overhead. About sunset they halt, and uttering low notes, dive into the thickets to feed, and afterward to rest. They go singly, or two or three together. The males come in advance, and instantly announce their presence by a loud and joyous song, continually emulating one another during the week or more that elapses before the arrival of the females. But this emulation does not end with vying in song; they have many pitched battles, chasing each other from tree to tree and through the branches with angry notes. The coming of the females offers some diversion to these pugnacious cavaliers, or at least furnishes a new *casus belli;* for, while they devote themselves with great ardor to wooing and winning their coy mistresses, their jealousy is easily aroused, and their fighting is often resumed. Even the lady-loves sometimes forget themselves so far as to savagely attack their fancied rivals, or drive out of sight the chosen mate of some male bird whom they want for themselves. This is not all fancy, but lamentable fact.

Mademoiselle Oriole is not so showy as her gay beau. Persuade the pair to keep quiet a moment, and compare them. They are in size between a bluebird and a robin, but rather more slender than either. The plumage of the male is of a rich but varying orange upon all the lower parts, underneath the wings, upon the lower part of the back, and the outer edges of the tail; the throat, head, neck, the part between the shoulders, wing quills, and middle tail feathers are velvety black; the bill and feet are bluish; there is a white ring around the eye, and the lesser wing quills are edged with white. In the female the pattern of color is the same, but the tints are duller. The jet of the male's head and neck is rusty in his mate, and each feather is margined with olive. The orange part of the plumage is more like yellow in the female, and wing and tail quills are spotted and dirty. Three years are required for the orioles to receive their complete plumage, the gradual change of which is beautifully represented in one of Audubon's gigantic plates. "Sometimes the whole tail of a [young] male individual in spring is yellow, sometimes only the two middle feathers are black, and frequently the black on the back is skirted with orange, and the tail tipped with the same color." Much confusion arose among the earlier naturalists from this circumstance, though not quite so much as ensued upon the discovery of the cousin of this species—the orchard oriole—which bears the specific name *spurius* to this day as a memory of the time when ornithologists called it a "bastard."

The singing of the males is at its height now that the females have come, and they are to be heard, not only from field and grove and country way-side, but in the streets of villages, and even in the parks of cities, where they are recognized by every school-boy, who calls them fire-birds, golden-robins, hang-nests, and Baltimore birds. The lindened avenues of Philadelphia, the elm-embowered precincts of New Haven, the sacred trees of Boston Common, the classic shades of Harvard Square, and the malls of Central Park all echo to their spring-time music.

The song of the oriole is indescribable, as to me are the tunes of most of the songsters. Nuttall's ingenious syllables are totally useless in expressing the pure and versatile fluting which floats down from the elm top. Wilson catches its spirit when he says that "there is in it a certain wild plaintiveness and *naïveté* extremely interesting," and that it is uttered "with the pleasing tranquillity of a careless ploughboy whistling for his own amusement." It is a joyous, contented song, standing out from the chorus that greets our half-awakened ears at daylight as brightly as its author shines against the dewy foliage. T. W. Higginson exclaims, "Yonder oriole fills with light and melody the thousand branches of a neighborhood." It is a song varying with the tune and circumstances, and, as among all birds, some orioles are better performers than others. Dr. Brewer thought that when they first arrived, and were awaiting the females, the voices of the males were loud and somewhat shrill, as though in lamentation, and that this song changed into a "richer, lower, and more pleasing refrain" when they were joined by their partners. The quality of their music is certainly different in different parts of the country, seeming, for example, to be more subdued toward the northern limit of their range.

A writer in an old number of *Putnam's Magazine* describes two orioles with which he had been acquainted for several summers. These birds had taken up their residences within about a quarter of a mile of each other, one in a public park, and the other in an orchard. "And often," says the narrator, "have I heard the chief musician of the orchard, on the topmost bough of an ancient apple-tree, sing,

to which the chorister of the park, from the summit of a maple, would respond, in the same key,

and, for the life of me, I never was able to tell whether their songs were those of rivalry or of greeting and friendly intercourse. And now if you will strike these notes on the piano, or, which is better, breathe them from the flute, you will know the song of the oriole, or rather obtain an idea of its general characteristics, for no two that I have ever heard sung the same melody."

The female also has a pretty song, which mingles with the brilliant tenor of the male during all the season of love-making; but as May merges into June, and the business of the summer begins, both cease their exalted strains, and only the mellow, ringing whistle is heard; then, as family cares increase, they lay aside even this, and, except at dawn, are rarely heard at all.

But, after all, the chief interest about our oriole is its wonderful home, which hangs upon the outmost branches of the elms along the street or in the grove, and is completed by June 10. The nest is never found in the deep woods. Its maker is a bird of the sunlight, and is sociable with man. The haunts of the orioles are those grand trees which the farmer leaves here and there in his field as shade for his cattle, to lean over the brier-tangled fence of the lane, or droop toward the dancing waters of some rural river. "There is," says Thomas Nuttall, "nothing more remarkable in the whole instinct of our golden-robin than the ingenuity displayed in the fabrication of its nest, which is, in fact, a pendulous, cylindric pouch of five to seven inches in depth, usually suspended from near the extremities of the high drooping branches of trees (such as the elm, the pear, or apple tree, wild-cherry, weeping-willow, tulip-tree, or button-wood)."

These words might in a general way apply to all the *Icteri*, most of which inhabit North or South America, have brilliant plumages, and build nests of matchless workmanship, woven and entwined in such a way as would defy the skill of the most expert seamstress, and unite dryness, safety, and warmth. They are mostly pendulous from the ends of branches, and form thus a security from snakes and other robbers, which could easily reach them if placed on a more solid foundation. They are formed of the different grasses, dry roots, lichens, long and slender mosses, and other advantageous materials often supplied by man's art. Among different species the structures vary in shape from resembling a compact ball to nearly every bottle-shaped gradation of form, until they exceed three or four feet in length. Many species being gregarious, they breed numerously in the same vicinity or on the same tree, resembling in this and other respects the weaver-birds, to which they are closely allied. But for us our Baltimore's nest possesses the most attractions; and as I shall have much to say concerning this fine example of a bird's architecture, I cannot begin better than by quoting Nuttall's description of it. It would be impossible for me to say anything different, and as well:

"It is begun by firmly fastening natural strings of the flax of the silk-weed, or swamp hollyhock, or stout artificial threads, around two or more forked twigs, corresponding to the intended width and depth of the nest. With the same materials, willow down, or any accidental ravellings, strings, thread, sewing-silk, tow, or wool that may be lying near the neighboring houses or around grafts of trees, they interweave and fabricate a sort of coarse cloth into the form intended, toward the bottom of which they place the real nest, made chiefly of lint, wiry grass, horse and cow hair:

sometimes, in defect of hair, lining the interior with a mixture of slender strips of smooth vine bark, and rarely with a few feathers; the whole being of a considerable thickness and more or less attached to the external pouch. Over the top the leaves, as they grow out, form a verdant and agreeable canopy, defending the young from the sun and rain. There is sometimes a considerable difference in the manufacture of these nests, as well as in the materials which enter into their composition. Both sexes seem to be equally adepts at this sort of labor; and I have seen the female alone perform the whole without any assistance, and the male also complete this laborious task nearly without the aid of his consort, who, however, in general, is the principal worker."

Many persons believe that there is a constant tendency in birds to vary their architecture to suit their surroundings, in accordance with climate, greater or less readiness of certain materials, and security. The Baltimore oriole affords a good illustration of this tendency. Like the swallows, robin, bluebird, pewit, and others, the oriole has abandoned the wilds for the proximity to man's settlements, doing it chiefly for two reasons—the greater abundance of insect food, and protection from hawks, owls, and crows, which are fewer in number and less bold in the clearings.

In the swamps of the Gulf States, the Baltimore, finding no necessity for great warmth or shelter from chilling winds, fabricates an airy nest of Spanish moss (*Tillandsia usneoides*). Audubon described and figured such a one, but the exact truth of Audubon's description was rather doubted until the Boston Society of Natural History received other similar nests from Florida. In these cases the bird chose material perfectly suited to the temperature, in preference to the flax and felt which it would have used in the North. This is a modification due to difference of latitude and accompanying difference of climate; but I venture to say that the Baltimores' nests, in general, built during an unusually hot season in any latitude will be much lighter than those built during a cool or backward year.

We may suppose that the oriole, having learned that the place for its home safest from all marauding animals and reptiles was out upon the tips of the swaying twigs, which would not bear the marauder's weight, would also have learned the shape best adapted to that situation; and that if it knew enough to choose the lesser danger from man in order to escape a greater one from hawks when it came out of the deep woods, it would also have reason enough to alter its style of building in such a way as should best hide the sitting bird from the prying eyes of its winged enemies, and at the same time afford dryness and warmth to the interior. Both of these were secured in the thick branches of the primeval forest by the leaves overhead and around. It is hence found that in the same climate the more exposed a nest is the denser its composition, the deeper the pouch, and the smaller its mouth. Pennant and others of the earlier writers on American birds described the orioles' nests as having only a hole near the top for entrance and exit, like those of some of the South American species. Wilson, who was the first real critic of our ornithology, said this was certainly an error, adding, "I have never met with anything of the kind." Both authors seem to have made too sweeping assertions, and, as usual, there is a golden mean of fact. Our hang-nest has enough discernment to select the safest and best site for a nest ever chosen by a tree-building bird. He has sufficient discretion to inhabit trees where his young will be least exposed to birds of prey. He has sense and skill enough to build a warm or cool house to suit the climate—a deep and tight one where the sun shines brightly, and sharp eyes might see the orange coat of himself or his mate within, and a loose and (in labor) less expensive one where deep shadows hide it. Surely, then, this consummate workman has ingenuity enough to put a roof over his dwelling to shed the rain and the hawk's glances, leaving only a little door in the side. Both of these things the hang-nest actually does. I myself have seen a nest of his making, over the open top of which a broad leaf had been bent down and tied by glutinous threads in such a way as to make a good portico. Mr. Thomas Gentry found a much more complete example at Germantown (Philadelphia), Pennsylvania, where the orioles " were constrained to erect a permanent roof to their dwelling by interwoven strings through their deprivation of the verdant and agreeable canopy which the leaves would naturally afford..... So nicely is the roof adjusted that even the most critical investigation cannot discern the union. The entrance is a circular opening situated in the superior third of the nest, facing southwardly." Mr. Gentry considers this the latest improvement upon a nest which in the beginning was simply a hammock in the fork of a tree, like a vireo's, but which has been made more and more pendulous, until what was at first the whole nest is now only the lining at the bottom of a deep enclosing bag.

With the idea of testing Wallace's theory that birds of bright colors, easily detected by birds of prey, are always found to occupy concealing nests, Dr. C. C. Abbott, of Trenton, New Jersey, made extensive notes upon the nests of our subject. In every instance those nests which fully concealed the sitting bird were at a considerable distance from any house in uncultivated parts. In all such localities sparrow-hawks were seen frequently, as compared with the neighborhoods selected for building the shallower open-topped nests, all of which were in willow or elm trees in the yards of farm-houses. The conclusion drawn was that the orioles knew where danger from hawks was to be apprehended, and constructed accordingly—the less elaborate nest in the farmer's yard answering every purpose for incubation. Dr. Abbott says, however, that of the nests that did conceal the sitting bird, every one was really open at the top, and the bird entered from above. Its weight, when in the nest, appeared to draw the edges of the rim together sufficiently to shut out all view of the occupant. It is his opinion, however, that years ago, when its enemies were more numerous, the nest of this oriole was perfectly closed at the top, and with a side opening; but he finds none so now.

The question why this species alone among our birds is supposed to have learned by dear experience to take such precautions against its foes has already been answered: it is because the Baltimore oriole is almost the only species in which the female is not protected from observation by her neutral and dull colors, and in which the brightly plumaged male also sits upon the eggs. Mother Necessity has prompted the marvellous invention.

Nuttall thought both sexes equally expert at nest-building, although the labor principally devolved upon the female. The latter clause in particular Mr. Gentry has confirmed, and tells us that the male occupies himself only in collecting materials for his mate. They labor very steadily, but a week's work is necessary for the completion of their home. It seems strange that domiciles constructed with so much pains should not be occupied successive seasons, but this seems never to be the case. It sometimes happens, however, that orioles will pick to pieces an old nest to get materials for a new one, just as the Indians of Peru often construct their huts of the cut-stone blocks of the ancient palaces of the Incas. These birds are very knowing in gathering stuff for the framework of their homes, and perceive the adaptability to their needs of the housewife's yarn and laces, hung out to dry, much sooner than they perceive the immorality of stealing them. White cotton strings are rarely absent·from their nests, which are sometimes almost entirely composed of them. Some curious anecdotes have been related of this economical propensity and its results; Nuttall tells the following:

"A female (oriole), which I observed attentively, carried

off to her nest a piece of lamp-wick ten or twelve feet long. This long string and many other shorter ones were left hanging out for about a week before both ends were wattled into the sides of the nest. Some other little birds, making use of similar materials, at times twitched these flowing ends, and generally brought out the busy Baltimore from her occupation in great anger."

A lady once told John Burroughs that one of these birds snatched a skein of yarn from her window-sill, and made off with it to its half-finished nest. But the perverse yarn caught fast in the branches, and, in the bird's efforts to extricate it, got hopelessly tangled. She tugged away at it all day, but was finally obliged to content herself with a few detached portions. The fluttering strings were an eyesore to her ever after, and passing and repassing she would give them a spiteful jerk, as much as to say, "There is that confounded yarn that gave me so much trouble!"

A gentleman in Pennsylvania, observing an oriole beginning to build, hung out "skeins of many-colored zephyr yarn, which the eager artist readily appropriated. He managed it so that the bird used nearly equal quantities of various high, bright colors. The nest was made unusually deep and capacious, and it may be questioned if such a thing of beauty was ever before woven by the cunning of a bird."

The nest being done, the female begins to deposit her eggs on the successive day, and continues laying one each day until four or five are laid. The eggs are pointed oval, 0.90 by 0.60 of an inch in dimensions, grayish-white, with a roseate tinge in fresh and transparent specimens, and variously marked with blotches and irregular lines, like pen scratches, of purplish-brown. On the day following, incubation begins, and the eggs hatch at the end of about fifteen days, bringing it to the middle of June.

The courage and devotion of the parents in defence of their nests are known to every ornithologist. They expose themselves fearlessly to danger rather than desert their charge, and call upon heaven and earth to witness their persecution. I remember one such instance. I discovered a nest with eggs in a sycamore on the banks of the Yantic River, in Connecticut. In trying to examine it I roused the ire of the owners, who showed the most intense anger and dismay. Enjoying this little exhibition, I did all I could to terrify the fond parents without harming them at all, and then quietly watched the result. The birds flew close about the nest, screaming and uttering a loud rolling cry like a policeman's rattle, which very soon brought plenty of sympathetic and curious friends. A cat-bird ventured too near, and was pounced upon by the Baltimore with a fierceness not to be resisted. But when the cat-bird found he was not pursued beyond the shade of the tree, he perched upon a neighboring post, and by hissing, strutting up and down, and every provoking gesture known to birds, challenged the oriole, who paid no attention to his empty braggadocio. Next Mrs. Oriole did something distasteful to her lord, and received prompt chastisement. A confident kingbird dashed up, and was beautifully whipped in half a minute. Vireos, pewits, warblers were attracted to the scene, but kept at a safe distance. There was no appeasing the anxiety of the parents until I left, and probably they spent the whole afternoon in recovering their equanimity.

The study of the expressions and dialects of animals and birds under such circumstances is extremely entertaining and instructive. Though you should happen upon a Baltimore's nest when the female is sitting, and the male is out of sight, the female will sit quietly until the very last moment; and Mr. Ridgway mentions an instance where the female even entered her nest while he was severing it from the branch, and remained there until carried into the house. The young birds, before they can fly, Dr. Brewer says, climb to the edge of the nest, and are liable in sudden tempests to be thrown out. If uninjured they are good climbers, and by means of wings, bill, and claws are often able to reach places of safety. In one instance a fledgling which had broken both legs, and had been placed in a basket to be fed by its parents, managed by wings and bill to raise itself to the rim, and in a few days took its departure. To this dexterity in the use of the bill as a prehensile organ, the birds may owe their skill in weaving.

The young are fed upon an insect diet, and mainly upon caterpillars which are disgorged after having been properly swallowed by the parents. They leave the nest after a fortnight, but are attended by the parent birds ten days longer before being turned off to take care of themselves. The food of the Baltimore oriole, old and young, is almost entirely insectivorous, succulent young peas and the stamens of cherry and plum flowers forming the only exceptions. These small robberies are but a slight compensation for the invaluable services he renders the gardener in the destruction of hosts of noxious insects. At first beetles and hymenopterous insects form his diet, and he seeks them with restless agility among the opening buds. As the season progresses, and the caterpillars begin to appear, he forsakes the tough beetle, and rejoices in their juicy bodies, being almost the only bird that will eat the hairy and disgusting tent-caterpillar of the apple-trees.

About the middle of September the Baltimore orioles begin to disappear, and by the last of the month all have left the Northern States for their winter-quarters in Mexico, Central America, and the West Indies.

XI.

BANK-SWALLOWS.

THE bird which is the subject of this sketch is familiar to all who walk in green pastures and beside still waters; for in such haunts do the bank-swallows congregate in merry companies, making up for their want of companionship with man, which is so characteristic of the other hirundines, by a large sociability among themselves. Conservator of ancient ways, it is almost the only swallow which has not attached itself to humanity as soon as it had opportunity, and changed from a savage to a civilized bird. Perhaps it, too, has tried it long ago, and voluntarily returned to the fields; for our bank-swallow is a cosmopolite, and has watched the rise and fall of all the dynasties and nationalities that have grouped the centuries into eras, from Nineveh to San Francisco. Even now it is an inhabitant of all Europe and eastward to China; of a large part of Africa, especially in winter; and throughout North America, the West Indies, Central America, and the northern

Andean countries. On both continents its wanderings extend to the extreme north, where, in Alaska, it is one of the commonest summer visitors. So this modest little bird, smallest of his kind, is entitled to our respect as a traveller at least; and, to compare the habits and appearance of the representatives in different portions of the globe of so widely distributed a species, becomes a most interesting study.

Cotyle riparia, the bank-swallow, sand-martin, sand-swallow, river-swallow, l'hirondelle de rivage, or back-svala, is generally diffused over the northern hemisphere, though very unequally, avoiding those spots unfavorable to it. In this distribution it seems to have been somewhat influenced by man, though owing him no other favors than the incidental help of railroad-cuttings and sand-pits, which have increased the sites suitable for its nests, and thus enabled the species to spread inland.

It is one of the earliest birds to arrive in the spring, appearing in Old England during the last week in March, and in New England early in May — many passing on to the shores of the Arctic Ocean, where Richardson, at the mouth of the Mackenzie, and Dall, on the Yukon, found them breeding in immense numbers. In these high latitudes its summer is necessarily a brief one, and September finds

FUN FOR THE BOYS, BUT—

it back again, picking up congeners for company on the southward journey.

Where these and other swallows spend the winter was a hotly-debated question among ornithologists at the beginning of the present century; some affirming that they migrated with the sun, while others, believing it impossible that such small and delicate birds could endure the great fatigue and temperatures incident to such a migration, held that they regularly hibernated during the cold weather, sinking into the mud at the bottom of ponds, like frogs, or curling up in deep, warm crannies, like bats, and remaining torpid until revived by the warmth of spring. Of this latter opinion was White, of Selborne, who alludes to it again and again; and Sir Thomas Forster wrote a "Monograph of British Swallows," apparently with no other object than to present the arguments for and against the theory of their annual submersion and torpidity. One of the difficulties which the *submersionists* put in the way of the *migrationists* was the frequent accidental and isolated appearance of the swallow before its usual time—a fact which has occasioned a proverb in almost every language. The French have, "*Une hirondelle ne fait pas le printemps;*" the Germans, "*Eine Schwalbe macht keinen Sommer;*" the Dutch, "*Een zwaluw maak geen zomer;*" the Italians, "*Una rodine *non fa primavera;*" the Swedes, "*En svala gor ingen sommar;*" which all mean, *One swallow doth not make a summer.* The story is well known of a thin brass plate having been fixed on a swallow with this inscription: "Prithee, swallow, whither goest thou in winter?" The bird returned next spring with the answer subjoined, "To Anthony, of Athens. Why dost thou inquire?"

Out of this controversy evidence of their sudden autumnal adjournment to Africa accumulated in England. Wilson, in this country, showed that their advance could be traced in the spring from New Orleans to Lake Superior and back again, and their regular migration soon came to be acknowledged. Then attention was turned to the season, manner, and limits of their migrations, and it was found that, taking advantage of favorable winds, immense flocks of swallows — and many other birds of passage as well — flying very high, passed each fall from the coast of England to the coast of Africa, and from Continental Europe across the Mediterranean direct, whence they spread southward almost to the Cape of Good Hope. No sooner had the spring fairly opened than they were suddenly back again, very much exhausted at first with their long-sustained effort, but speedily recuperated and "diligent in business." Our own migrants, as I have mentioned, winter

A NARROW ESCAPE.

in Central America and the West Indies, or still farther south.

Their flight is rapid but unsteady, "with odd jerks and vacillations not unlike the motions of a butterfly," as White describes it; and continues: "Doubtless the flight of all hirundines is influenced by and adapted to the peculiar sort of insects which furnish their food. Hence it would be worth inquiry to examine what particular genus of insects affords the principal food of each respective species of swallow." They are constantly on the wing, skimming low over land and loch, pausing not even to drink or bathe, but simply dropping into some limpid lake as they sweep by to sip a taste of water or cleanse their dirty coats. It seems strange, then, that birds who sustain the unremitting exertion of a flight scarcely less than one hundred miles an hour in speed, during the whole of a long summer's day, should not be thought capable of the transition from England to Africa. However, at that time it was not well understood what long-continued flight small birds actually do make, as, for instance, from our coast to the Bahamas, or even across to Ireland, or from Egypt to Heligoland, one thousand two hundred miles, which is passed over at a single flight, by a certain tiny warbler, in every migration.

The bank-swallow is not a musical bird, a faint, squeaking chirrup being all its voice can accomplish; nor is it a handsome bird—simply sooty-brown above, white beneath, with a brown breast. To its grace of motion and charming home-life we attribute that in it which attracts us.

Although probably the least numerous of all the swallows, they do not seem so, because of the great companies which are to be seen together wherever they are to be found at all; and because, leading a more sequestered life, they are not usually brought into direct comparison with house-martins and chimney-swifts. Eminently social in their habits, they congregrate not only at the time of migration (then, indeed, least of all), and in the construction of their homes, but sometimes alight in great flocks on the reeds by the river-side and on the beach, where Sir William Jardine saw them, "partly resting and washing, and partly feeding on a small fly, which was very abundant." Yet you will occasionally notice stray individuals associating with other swallows.

The secret of the local distribution of the bank-swallows lies in the presence or absence of vertical exposures of soil suitable for them to penetrate for the burrows at the inner end of which the nest is placed. Firm sand, with no admixture of pebbles, is preferred, and in such an exposure, be it sea-shore, river-bank, sand-pit, or railway-cutting, the face will be fairly honey-combed with burrows, so that we can readily believe that Mr. Dall counted over seven hundred holes in one bluff in Alaska. These are usually very close together, and the wonder is how the birds can distinguish their own doors. If mistakes do occur, I imagine they are all very polite about it, for I know of no more peaceable neighbors among birds than they. The mode in which this perforation is performed, requiring an amount of labor rare with animals, is well described by Mr. Rennie in his "Architecture of Birds:"

"The beak is hard and sharp, and admirably adapted for digging; it is small, we admit, but its shortness adds to its strength, and the bird works..... with its bill shut. This fact our readers may verify by observing their operations early in the morning through an opera-glass, when they begin in the spring to form their excavations. In this way we have seen one of these birds cling with its sharp claws to the face of a sand-bank, and peg in its bill as a miner would his pickaxe, till it had loosened a considerable portion of the hard sand, and tumbled it down among the rubbish below. In these preliminary operations it never makes use of its claws for digging; indeed, it is impossible that it could, for they are indispensable in maintaining its position, at least when it is beginning its hole. We have further remarked that some of these martins' holes are nearly as circular as if they had been planned out with a pair of compasses, while others are more irregular in form; but this seems to depend more on the sand crumbling away than upon any deficiency in its original workmanship. The bird, in fact, always uses its own body to determine the proportions of the gallery, the part from the thigh to the head forming the radius of the circle. It does not trace this out as we should do, by fixing a point for the centre around which to draw the circumference; on the contrary, it perches on the circumference with its claws, and works with its bill from the centre outward;..... the bird consequently assumes all positions while at work in the interior, hanging from the roof of the gallery with its back downward as often as standing on the floor. We have more than once, indeed, seen a bank-martin wheeling slowly round in this manner on the face of a sand-bank when it was just breaking ground to begin its gallery.

"This manner of working, however, from the circumference to the centre unavoidably leads to irregularities in the direction...... Accordingly, all the galleries are found to be more or less tortuous to their termination, which is at the depth of from two to three feet, where a bed of loose hay and a few of the smaller breast-feathers of geese, ducks, or fowls is spread with little art for the reception of the four to six white eggs. It may not be unimportant to remark, also, that it always scrapes out with its feet the sand detached by the bill; but so carefully is this performed that it never scratches up the unmined sand, or disturbs the plane of the floor, which rather slopes upward, and of course the lodgment of rain is thereby prevented."

Sometimes the nest is carried to a far greater depth than two or three feet, as in a case observed by Mr. Fowler, in Beverly, Massachusetts, where, in order to get free of a stony soil, where pebbles might be dislodged and crush the eggs, the tunnel was carried in nine feet, while neighboring birds in better soil only went a third as far. In one place the burrows will be close to the top of the bluff; in another near the bottom, according as fancy dictates, or the birds have reason to fear this or that enemy. English writers agree that occasionally their bank-swallows do not dig holes, but lay in the crannies of old walls, and in hollows of trees. This is never done, that I am aware of, in the United States; but in California a closely allied species, the rough-winged swallow, "sometimes resorts to natural clefts in the banks or adobe buildings, and occasionally to knot-holes." On the great plains, however, our *Cotyle* burrows in the slight embankments thrown up for a railway-bed, in lieu of a better place; and at St. Paul I have seen them penetrating solid, but soft, sand-rock.

"How long does it take the bird to dig his cavern under ordinary circumstances?" is a question which it would seem hard to answer, considering the cryptic character of his work. Mr. W. H. Dall says four days suffice to excavate the nest. Mr. Morris, a close observer of British birds, says, *per contra*, a fortnight; and that the bird removes twenty ounces of sand a day. Male and female alternate in the labor of digging, and in the duties of incubation.

When the female is sitting you may thrust your arm in and grasp her, and, notwithstanding the noise and violence attending the enlargement of the aperture of her nest-hole, she will sit resolutely on, and allow herself to be taken in the hand with scarcely a struggle or sign of resistance— even of life, sometimes. The young are fed with large insects caught by the parents, particularly those sub-aquatic sorts which hover near the surface of still water; and White mentions instances where young swallows were fed with dragon-flies nearly as long as themselves. The young do not leave the nest until they are about ready to take full care of themselves. Finally, they are pushed off by the parents to make way for a second brood, and, inexperienced in the use of their wings, many fall a prey to crows and small hawks that lie in wait ready to pounce upon the first poor little fellow that launches upon the untried air. Those that manage to run the gauntlet of the hawks collect in small companies by themselves, and have a good time hunting by day and roosting at night among the river-reeds, until the autumn migration. "At this time, Salerne observes," says Latham, "that the young are very fat, and in flavor scarcely inferior to the *ortolan*." Sometimes the parents forsake their progeny in the nest, and seem generally to care less for them than is usually the case among swallows.

But not the young alone are exposed to enemies. It would seem as though the situation of the nest precluded invasion, yet, if they are near the haunts of the house-sparrow, they are sure to be dispossessed of their homes by that buccaneer. Snakes, too, can sometimes reach their holes; weasels, like that one Mr. Hewitson tells us of, are often sharp enough to make their *entrée* from above: school-boys regard the pink-white eggs a fine prize; and, last and worst of all, the bank-swallows are many times utterly worried out of their galleries by fleas and young horse-flies, which swarm and increase in their nests until the bird finds endurance no longer a virtue, and digs a new *latebra*.

INDEX.

A.
Animals, comprehension of, 218.
" lost, returning, 199.

B.
BANK-SWALLOWS, 241.
Bee, tactics of, 203.
Birds and weather, 113.
" at night, 111.
" enemies of, 185, 255.
" far eyesight of, 213.
" flight of, 87, 249.
" food in winter, 131.
" migration of, 93, 107, 249.
" nests of, 102, 230, 249.
" of Spring, 36.
" OUR WINTER, 106.
" protective colors of, 134, 233.
" superstitions about, 196.
Bison (Buffalo), domesticated, 154.
" extinct races of, 143, 162, 164.
" former range of, 158.
Blackbird, the crow, 52.
" the red-winged, 49.

Bluebird, 37.
BUFFALO, THE, AND HIS FATE, 140.

C.
Chickadee, the, 122.
Chippy, the, 46.
CIVILIZING INFLUENCES (on birds), 182, 230, 241.
Civilizing influences on the bison, 166.
Creeper, the brown, 119, 135.

D.
Darwinism in the song-sparrow, 181.
Dogs, anecdotes of homing ability of, 207.
Dogs, power of scent in, 215.
Dove, the, 44.

F.
FIRST-COMERS, 36.
Fishes, possible perceptions of, 204.
Fringillidæ, the winter, 123, 136.

INDEX.

G.
Geographical distribution of birds, 91.
Geographical distribution of snails, 25.
Goldfinch, the, 125.
Grass-finch, the, 127.

H.
HOW ANIMALS GET HOME, 199.

J.
Julia Brace, sense of smell in, 217.

K.
Kinglets, the, 118, 134.

M.
Mice, damage done by, 82.
Mouse in vireo's nest, 73.
" jumping, 60.
" meadow, 63, 70, 79.
" white-footed, 74, 84.
Mound-builders, the, 163.
Mules, keen intelligence of, 205.

N.
Nuthatch, the, 120.

O.
Oriole, Baltimore, 220.
ORNITHOLOGICAL LECTURE, AN, 85.

P.
Pigeons, feats of homing, 199.
Pine-finch, the, 124, 136.
Pine-grossbeak, the, 124, 136.
PRINCE, A MIDSUMMER, 220.

R.
Red-poll linnet, the, 126.
Redwing, the, 49.

S.
SNAILERY, IN A, 9.
Snails, anatomy, 10, 21.
" as food and medicine, 32.
" eggs of, 15.
" hibernation of, 22.
" vitality of, 30.
" where to search for, 19.
Sparrow, chipping, 46.
" field, 128.
" THE SONG, 127, 170.
Swallows and swifts, 189.
" bank, 241.
Shrikes, 80, 137.
Snow-bird, the, 115.
Snow-bunting, the, 117, 136.

T.
Titmice, 122.

W.
Wren, common house, 41.
" winter, 123, 135.
WILD MICE, 57.

INTERESTING WORKS

ON

NATURAL HISTORY.

Biart's Adventures of a Young Naturalist.
    The Adventures of a Young Naturalist. By LUCIEN BIART. Edited and Adapted by PARKER GILLMORE. With 117 Illustrations. 12mo, Cloth, $1 75.

Darwin's Voyage of a Naturalist.
    Voyage of a Naturalist. Journal of Researches into the Natural History and Geology of the Countries visited during the Voyage of H.M.S. *Beagle* round the World, under the Command of Captain Fitzroy, R.N. By CHARLES DARWIN, M.A., F.R.S. 2 vols., 12mo, Cloth, $2 00.

What Mr. Darwin Saw.
    What Mr. Darwin Saw in his Voyage round the World in the Ship *Beagle*. With Illustrations. 8vo, Cloth, $3 00.

Gillmore's Prairie and Forest.
    Prairie and Forest: a Description of the Game of North America, with Personal Adventures in their Pursuit. By PARKER GILLMORE. Illustrated. 12mo, Cloth, $1 50.

Greenwood's Wild Sports of the World.
    Wild Sports of the World: a Book of Natural History and Adventure. By JAMES GREENWOOD. Illustrated. Crown, 8vo, Cloth, $2 50.

Interesting Works on Natural History.

Hooker's Natural History.

Natural History. For the Use of Schools and Families. By WORTHINGTON HOOKER, M.D. Nearly 300 Illustrations. 12mo, Half Leather, $1 00. *Uniform with the "Science for the School and Family" Series.*

Pike's Sub-Tropical Rambles.

Sub-Tropical Rambles in the Land of the Aphanapteryx: Personal Experiences, Adventures, and Wanderings in and about the Island of Mauritius. By NICHOLAS PIKE. Handsomely Illustrated. 8vo, Cloth, $3 50.

Jaeger's North American Insects.

The North American Insects, with Numerous Illustrations drawn from Specimens in the Cabinet of the Author. By Professor JAEGER, assisted by H. C. PRESTON, M.D. 12mo, Cloth, $1 50.

Kingsley's West Indies.

At Last: a Christmas in the West Indies. By CHARLES KINGSLEY. Illustrated. 12mo, Cloth, $1 50.

Lewes's Studies in Animal Life.

Studies in Animal Life. By GEO. H. LEWES. Illustrated. 12mo, Cloth, $1 00.

Smiles's Scotch Naturalist.

Life of a Scotch Naturalist: Thomas Edward, Associate of the Linnæan Society. By SAMUEL SMILES. Portrait and Illustrations. 12mo, Cloth, $1 50.

Interesting Works on Natural History. 3

Spry's Cruise of the "Challenger."

The Cruise of Her Majesty's Ship "Challenger." Voyages over many Seas, Scenes in many Lands. By W. J. J. SPRY, R.N. With Map and Illustrations. Crown 8vo, Cloth, $2 00.

Thomson's Voyage of the "Challenger."

The Voyage of the "Challenger." *The Atlantic:* An Account of the General Results of the Voyage during the Year 1873 and the Early Part of the Year 1876. By Sir C. WYVILLE THOMSON, F.R.S. With a Portrait of the Author, Colored Maps, Temperature Charts, and Illustrations. 2 vols., 8vo, Cloth, $12 00.

Wallace's Island Life.

Island Life; or, The Phenomena of Insular Faunas and Floras, with their Causes. Including an entire revision of the Problem of Geological Climates. By ALFRED RUSSEL WALLACE. With Illustrations and Maps. 8vo, Cloth. (*In Press.*)

Wallace's Malay Archipelago.

The Malay Archipelago: the Land of the Orang-Utan and the Bird of Paradise. A Narrative of Travel, 1854–'62. With Studies of Man and Nature. By ALFRED RUSSEL WALLACE. With Maps and numerous Illustrations. Crown 8vo, Cloth, $2 50.

Wallace's Geographical Distribution of Animals.

The Geographical Distribution of Animals. With a Study of the Relations of Living and Extinct Faunas, as Elucidating the Past Changes of the Earth's Surface. By ALFRED RUSSEL WALLACE. With Colored Maps and numerous Illustrations. 2 vols., 8vo, Cloth, $10 00.

4   *Interesting Works on Natural History.*

Treat's Chapters on Ants.
Chapters on Ants. By MARY TREAT. 32mo, Paper, 20 cents.

White's Selborne.
Natural History of Selborne. By Rev. GILBERT WHITE. Illustrated. 18mo, Cloth, 75 cents.

Wood's Homes without Hands.
Homes without Hands: being a Description of the Habitations of Animals, classed according to their Principle of Construction. By the Rev. J. G. WOOD, M.A., F.L.S. With about 140 Illustrations. 8vo, Cloth, $4 50; Sheep, $5 00; Roan, $5 00; Half Calf, $6 75.

Wood's Man and Beast, Here and Hereafter.
Man and Beast, Here and Hereafter. Illustrated by more than Three Hundred Original Anecdotes. By the Rev. J. G. WOOD, M.A., F.L.S. 8vo, Cloth, $1 50.

Wood's The Illustrated Natural History.
The Illustrated Natural History. By the Rev. J. G. WOOD, M.A., F.L.S. With 450 Engravings. 12mo, Cloth, $1 05.

PUBLISHED BY HARPER & BROTHERS, NEW YORK.

☞ HARPER & BROTHERS *will send the above works by mail, postage prepaid, to any part of the United States, on receipt of the price.*

www.ingramcontent.com/pod-product-compliance
Lightning Source LLC
Chambersburg PA
CBHW021344230426
43666CB00006B/397